A COMPREHENSIVE GUIDE TO LAND NAVIGATION WITH GPS

BY
NOEL J. HOTCHKISS

PUBLISHED BY:

ALEXIS U.S.A.

1037 Sterling Road • Suite 201
Herndon, Virginia 22070

First Printing 1994
Second Printing 1995 Revised

Published by:
Alexis Publishing
A Division of ALEXIS USA, INC.
1037 Sterling Road • Suite 201
Herndon, Virginia 22070

**Library of Congress Catalog
Card Number: 95-80632**

ISBN No: 0-9641273-3-4

Printed in Hong Kong through Mandarin Offset.

GRAND CANYON, W. H. HOLMES CIRCA 1882

ABOUT THIS GUIDE

This practical, hands-on guide was written to meet the needs of those who wish to **apply** the advantages of the new Global Positioning System (GPS) technology to their navigational requirements on land. Unlike many publications on the subject, this one does not dwell upon how the system works. Instead, it is focused on what GPS can do and how it can best be used by land navigators in a variety of circumstances.

Although this publication is written to address the needs of general and recreational consumers, it will also be of great interest to those who wish to explore the applicability of GPS to their professional land navigation (LN) requirements. This includes public safety, forestry and even military personnel.

This technology is new, so its utility is not yet fully understood by many who could take advantage of it. This is especially true for potential land-based users because they have had less experience with some of the techniques associated with using maps and charts, coordinate systems, and the more formalized navigational practices commonly employed in both flight and nautical situations. Nevertheless, this lack of familiarity should not deter anyone. This guide contains all the information needed to make navigating with GPS easy even for the novice.

In addition to the general lack of awareness people may have regarding GPS's great utility to the land-based navigator, there are some serious misconceptions that must be dispelled. First, there is the belief that GPS user equipment is too expensive to purchase and employ for general use by average consumers, state and local public safety agencies, or volunteer emergency organizations. The fact is that the prices of GPS receiver units have fallen dramatically since they were first introduced a few years ago.

Another widely held misconception is that consumers must purchase, install, and carry with them expensive and bulky electronic equipment for accessing and displaying digital or interactive videodisk maps in order to effectively employ their GPS units. While there are now available some complex and expensive integrated systems using various digitalized mapping data sources, such as the Geographic Information System (GIS), there is still a very attractive alternative available: paper maps.

For centuries people have been using good paper maps to help them find their way. There is no reason paper maps cannot continue to serve us well as we enter this new era of navigation with the GPS. Paper maps are inexpensive, readily obtainable, and easy to carry along in a vehicle or on foot. In fact, it is likely that the expanding use of GPS will spur both an interest in geography as well as dramatically increase the demand for paper maps everywhere.

In summary, adding the advantages of using GPS to your current repertoire of navigational skills and capabilities is the most convenient and cost-effective way to ensure that you will arrive at your destination quickly and without difficulty.

Finally, this Guide is organized in a way that will make it easy to use and understand.

Chapters 1 through 4 focus upon how to fully exploit and integrate GPS capabilities into a variety of navigational situations without delving into the specific instructions on how to operate the equipment. This discussion is reserved for Chapter 5. Learning how to employ the GPS receiver comes more logically after you know about its advantages and how it can and should be employed. The equipment, itself, is not difficult to operate. Chapter 6 provides a few suggestions for both developing and participating in a short practice navigation field exercise designed to apply the skills learned about LN while using the equipment. The final chapter (Chapter 7), provides a short glimpse into the future by forecasting some of the impact GPS will have upon how we do things.

In conclusion, Navstar, the Global Positioning System, was planned, designed, and developed by the United States Department of Defense as a navigational aid for military operations. But, without question, it offers great utility to both military and civilian consumers.

This Guide will introduce you to the unparalleled advantages and unexpected operational simplicity that is built into this revolutionary new GPS land navigation equipment.

After you have read the manual, we invite you to send your questions, reactions, and suggestions along to us at:

Alexis Publishing
1037 Sterling Road
Suite 201
Herndon, Virginia 22070

We will attempt to answer your questions and will consider your reactions and suggestions as we develop future publications and products.

TABLE OF CONTENTS

ILLUSTRATION BY MATTHEW DWYER

GPS OVERVIEW

All the great navigators - including Marco Polo, Magellan, Lewis & Clark, and you - have shared a common problem over the ages. Until recently, there was no surefire way to quickly and easily determine a position. Now, with GPS, this problem has been solved.

BACKGROUND

Always knowing your location is the fundamental key to finding your way along a selected route to your destination. It is the most fundamental requirement for any type of navigation.

Over the years, several aids and techniques were developed and used to help guide our movements. They included the drawing of crude portolan charts of the Mediterranean during the Middle Ages and use of the astrolabe, sextant, and printed almanacs that predict the "movements" of celestial bodies across our skies. More recently, we have employed modern radio-based systems, such as the Navy's low-orbit Transit SatNav (satellite navigation) and surface broadcasting concepts like Omega and Loran-C.

All of these aids had limitations associated with their use that are not encountered when using Global Positioning System (GPS) equipment. These limitations included poor visibility related to weather, the necessity to refer to endless tables, and the requirement to follow mathematically complex and time-consuming procedures. In the case of electronic navigation, the limitations were restrictive times and areas of coverage, increasingly questionable accuracies at greater ranges, and atmospheric and ground interferences that often resulted in poor signal reception. Low altitude and ground-based radio systems are particularly impractical on land.

NAVSTAR GPS, an acronym standing for **Nav**igation **S**ystem with **T**ime **A**nd **R**anging **G**lobal **P**ositioning **S**ystem, is a revolutionary development that is designed to provide highly accurate, reliable, continuous 24-hour, worldwide coverage for position reporting. It was created by the United States Department of Defense and is operated by the Air Force. The satellites that make up the space segment of the system broadcast the information required for the small lightweight user unit (receiver/computer) to determine its precise location. While the design of the system and the hand-held GPS receiever are highly complex, use of the equipment is simple. You can have your position reported on the display screen in seconds with the push of a button.

During the development and testing of the **NAVSTAR** program, the United States Government made decisions to extend its use to both domestic and international communities. Its applications range from navigation over the land, in the air, and on the seas to precision surveys, tracking the whereabouts of trains and trucks, and locating, for disposal, the mines left behind in Kuwait in the aftermath of the 1991 Gulf War. Our imaginations seem to be the only limits to the applications that may be developed for GPS. Of course, our discussion here will be limited to those closely related to its use as an aid to our movements over the land.

Some see GPS as a new man-made form of celestial navigation because a constellation of 24 artificial satellite "stars" (including three spares) has been placed in high orbit circling the globe every 12 hours at an altitude of approximately 12,500 statute miles (about 20,200 kilometers). Each satellite tells an unlimited number of GPS receivers anywhere in the world the current date, time, and locations of all other NAVSTAR satellites both now and into the immediate future. It also sends out a precise timing code. This timing code makes it possible for each navigational unit to calculate its distance from either three or four satellites presently "in view" and then determine its position.

Information from three satellites is needed to calculate a navigational unit's horizontal location on the earth's surface (2-dimensional reporting); but information from four satellites enables it to determine its altitude in addition to its horizontal location (3-dimensional reporting). Three-dimensional reporting is more crucial on land because, unlike the surface of a large body of water, ground surfaces are not constant and the elevation of the receiver antenna is considered when the unit calculates its horizontal position on the ground. Ninety-five percent of all position fixes are accurate to within 25 to 100 meters. This is close enough for nearly all land navigation applications. Without question, it will keep you from being lost.

This system could be even more precise if the U.S. Department of Defense were to drop its **Selective Availability Program (SA)** that makes GPS less accurate for security reasons. On the other hand, it is possible to overcome the SA and other inaccuracies through use of a special, yet somewhat expensive, application called **differential GPS**. This will be discussed later.

If you wish to further explore a layperson's discussion of the technical aspects of this new space-age technology, please refer to Appendix B in the back of the Guide.

GPS APPLICATIONS FOR LAND NAVIGATION (LN)

Many of you first learned about the GPS on TV or in newspaper stories reporting the progress of our allied forces during OPERATION DESERT STORM. Some of the popular hobbyist- and scientifically-oriented magazines then followed-up with short articles reflecting the enthusiasm expressed by our troops for this new navigational aid. As the result of the Gulf War, many soldiers discovered its advantages when they were issued (and some even purchased) this new equipment. More recently, we have seen news stories telling us about soldiers being trained in the use of GPS just prior to their deployment as part of OPERATION RESTORE HOPE in Somalia. And, finally, some of you may have friends or acquaintances interested or involved in the fields of nautical or aerial navigation who have shared some of the excitement and promise GPS is bringing to these audiences.

What is important is that you understand **GPS is for everyone** - not just the driver of a highly computerized tank, or jet pilot, or ship's captain. GPS has and certainly will continue to be integrated into a growing variety of advanced "heads-up" displays, digital computer and interactive laser videodisk map call-ups, and sophisticated nationwide vehicle location and tracking systems. But, without question, its largest user group will consist of individuals using relatively inexpensive nonintegrated GPS equipment while carrying paper maps of the areas in which they must find their way.

Next, we will briefly introduce some of the navigational advantages offered by

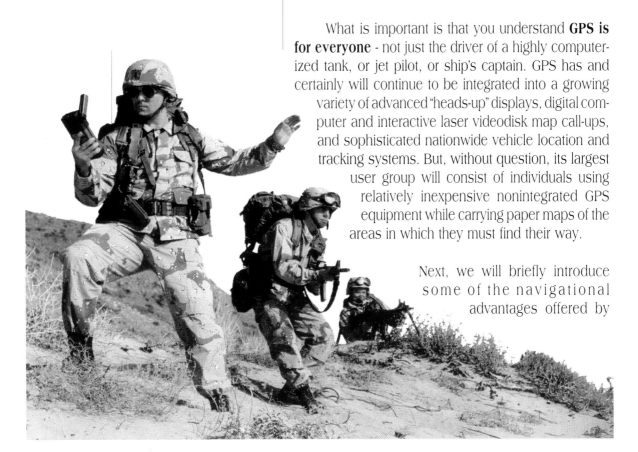

GPS receivers. Later chapters will help you to investigate and utilize these many useful functions.

FOR GENERAL NAVIGATIONAL USE

NAVSTAR GPS (the system) has only one fundamental purpose. It sends out radio signals carrying the information needed by your receiver unit to compute its position. It provides continuous, worldwide coverage under any atmospheric condition.

On the other hand, receiver units are designed to perform a multitude of functions. They are extremely versatile navigational tools.

For example, the GPS receiver not only determines and reports where you are located, it also remembers where you have been. Thus, it can tell you how to get back (direction and distance) to any location you have asked it to recall or that you may wish to punch in with a few strokes on the keypad.

Since it can tell you where you are and remembers where you were, it can also determine the **direction** you have gone and calculate how far you have moved during a given period and report your **average speed** over the ground. Finally, after you have established and programmed the route you wish to follow (intermediate checkpoints and final destination) and begin to navigate over it, the unit will provide you with a wealth of navigational information to assist you with your move-

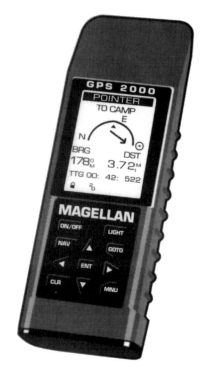

THE MAGELLAN GPS 2000 ™

ment. It will display, both in words and graphically, the name and relationship of your next waypoint to your present position and to north or south, its direction and distance, your actual movement direction, and your estimated time of arrival from your present position at any time along the way. In addition, it will report your velocity over the ground and the rate at which you are making progress toward your destination. Finally, it will show you which direction and how far you have drifted off your intended route and allow you to either get back on course or proceed directly to your next checkpoint.

How can the GPS receiver accomplish so much more for the land navigator than basic location reporting? Because, in addition to collecting, processing, and reporting positioning information from the NAVSTAR satellites, it can store and integrate this information with other data in various ways for later use. For example, the unit can draw upon and utilize a vast pool of pre-programmed formulas and reference tables held permanently in its memory, and it will accept a number and variety of information inputs and preferred reporting format selections made by you.

The pre-programmed reference information held within the unit's permanent memory is what makes it a comprehensive navigational aid. This information includes: (1) tables containing magnetic variations (declinations) for all parts of the world, (2) formulas for calculating and reporting coordinate locations on maps using any one of several geodetic datums (theorized earth surface shapes) upon which various map drawings are based, and (3) the tables and equations needed to report positions in either True Geographic Coordinates (Latitude and Longitude) or Universal Transverse Mercator Grid (UTM) coordinate systems. Either or both of these two standard coordinate systems are found on the majority of large and intermediate-scale maps produced around the world.

Chapter 3 will take an in-depth look at three of these standard coordinate systems. **Latitude and longitude** are universally used for reporting positions around the world and are found on most maps. Next, the **Universal Transverse Mercator Grid** (UTM) is printed on military maps of many areas (including the U.S.A.) and also included on all maps published by the U.S. Geological Survey (U.S.G.S.) and many other governments' mapping agencies. It forms a perpendicular grid with constant distances among its lines that makes it superior to latitude and longitude for use on land. And, finally, the **Military Grid Reference System** (MGRS) - not really a separate coordinate system at all - is a variation of the UTM grid system that makes it much easier to use. Whenever possible, use of the MGRS format of the UTM grid is the best bet for any type of land navigation.

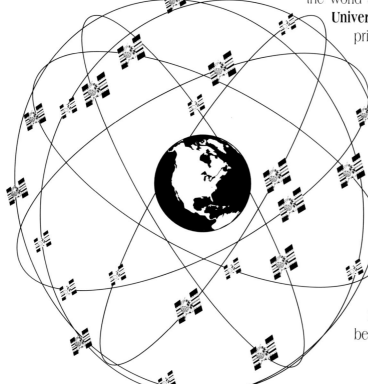

VECTOR ILLUSTRATION OF 24 SATELLITE GPS CONSTELLATION BY FRANCK E. BOYNTON

Presently, some GPS units do not report UTM positions in MGRS format, but many units do. For example, this includes Magellan's GPS 1000M5™ and the Trailblazer M™ as well as one of Trimble's Scouts, and other manufacturers' models.

The information inputs and reporting format options that you can program into most units allow you the flexibility to configure and re-configure it to meet your changing needs. For example, you may (1) **enter your initial position** to reduce the time it takes to obtain and to increase the accuracy of your first position-fix; (2) **save current and/or enter up to 100 positions to be used as waypoints** (checkpoints) for use in route planning; and (3) **program a multi-segment reversible route** with up to 15 legs that the unit will track in the navigation mode.

In addition, you can tailor the reports rendered by your receiver by making selections from among several available formats. They include: (1) **time and date information** (local or universal time), (2) **azimuth directions** (true, or magnetic values), (3) **distance, elevation, and speed readings** (metric or British units), and (4) **position-reporting modes** (3-D, 2-D, or automatic).

There are also several other functions and auxiliary features that make this equipment more accurate and easier and quicker to use. However, they will be reviewed in more detail in Chapter 5, where you will read how to specifically employ the many features found on the majority of GPS receiver units.

IN SUMMARY

If you are an experienced navigator, your thoughts may already be racing toward new applications for GPS. And, if you are not that experienced, this comprehensive Guide will help you to more fully explore and exploit the many navigational advantages it makes available to you.

It should be of little concern that you may not already understand such concepts as magnetic variation, map projections, or coordinate systems because these and many other aspects of using maps in conjunction with GPS will be fully explained in Chapter 3. In fact, the Guide will teach you all that is necessary about reading maps, observing the terrain, and using a magnetic compass to move over the land with skill and confidence while employing this revolutionary new equipment.

Whether you are an outdoorsman, traveler, part of a civilian public safety organization, forester, soldier, or anyone else wishing to improve your ability to navigate over the land, you will find both GPS and this guide quite helpful. Sometimes, improved capabilities in LN may be only a matter of convenience; but it is frequently essential for getting the job done and done safely. And, at times, it may even become a matter of life and death.

PHOTO: U.S. GEOLOGICAL SURVEY

BACKGROUND

THE UTILITY OF GPS ON LAND

Navigation is basically a problem-solving activity. It consists of a series of requirements or problems (many unforeseen) that are unique to every situation, each demanding the development of a strategy to meet or solve it. The purpose of this chapter is to show you how GPS technology and equipment can help you handle these challenges.

The fundamental concepts behind successful LN are not difficult to understand. In fact, there are only **four steps, one rule, and three movement techniques** to be employed as you find your way.

The four steps are;

Step 1 - **Know where you are,**

Step 2 - **Plan the route,**

Step 3 - **Stay on the route, and**

Step 4 - **Recognize the objective.**

The **rule** is that you should **always navigate with a correctly oriented map**. In fact, the amount of error should not exceed 30°. This practice of always keeping your map oriented as you make changes in direction is sometimes referred to as "steering" or "driving the map."

The three **movement techniques** that land navigators employ are **(1) terrain association, (2) dead reckoning, and (3) highway or trail following**. **Terrain association** consists of the navigator's use of the many features encountered in the real world and portrayed on maps to help determine position and guide movements. **Dead reckoning** (altered over the centuries through common language usage from deduced reckoning) is simply the application of an old nautical navigation technique to movements on land. The navigator uses a compass to follow a specific directional bearing and some means for measuring the distance traveled (e.g., an odometer, pacing, or lapsed time) to keep track of position and guide movement along each segment of the route. Finally, **highway / trail** following generally refers to a navigator's use of well established and mapped highway, street, and trail networks to track his position and guide him along to the destination. Obviously, these techniques can be employed separately or in concert.

GENERAL APPLICATION

Now it's time to take a closer look at the requirements and techniques associated with LN and how the Global Positioning System (GPS) can be applied in helping you to accomplish them. Let's first examine the four steps and three movement techniques.

FOUR-STEP LN PROCEDURE

STEP 1 (LOCATION AWARENESS)

Yes, you must continually know where you are located, both on the map and on the ground, in order to navigate effectively. This includes when you start your journey, while moving along the route, and as you reach your destination.

Before the advent of GPS, land navigators were required to almost constantly relate the features they saw in the real world around them with the map's portrayal in order to keep track of their locations. For even the most experienced navigators, there was always a large element of uncertainty for a number of reasons. First, no map presents a perfect portrait of the terrain. Due to scale limitations, maps are severely restricted as to how much detail they can include in a limited space without becoming an illegible mass of clutter. Next, there are always many changes made in the real world after any map is produced. This is particularly true with regard to man-made features and vegetation. And finally, mistakes are occasionally made by those who compile the maps we use.

Map interpretation, at best an imprecise science, can also present a difficult challenge for navigators. Each individual brings to the task different experiences and skill levels, and some terrain is more difficult to read than others. For example, the relatively featureless areas found along the borders of Saudi Arabia, Kuwait, and Iraq were nearly impossible for the allied Coalition Forces to use in determining their locations during Operations DESERT SHIELD and DESERT STORM.

All of these factors serve to erode any navigator's confidence during movements in those areas which are unfamiliar. Using maps, compasses, and protractors, military navigators and others who moved cross-country often completed time-consuming triangulations from two or more known features in the distance as a means to restore a degree of certainty and confidence in their position estimates.

Now, using GPS equipment, you can quickly learn your precise location anywhere at anytime. After dark or at midday, in rain or shine; or whether the surrounding countryside presents a featureless plain or numerous easily identified landmarks; the push of a button quickly calls-up a position reading that you can use with full confidence. That, in itself, is reason enough to employ this revolutionary new equipment, but there is much more.

STEP 2 (ROUTE PLANNING)

Land navigation's second step reminds you to plan the route. Let's imagine for a moment that you have decided to go hunting with your camera in search of wild game or some beautiful scenery. You pull the car over and stop along the edge of the logging trail you have been driving, jump out, and then push the button on your GPS unit requesting your current position. After obtaining that information, you store it as a waypoint named "CAR." Now, you move on quickly through the woods with your camera in hand. Several rolls of film and a few hours later, you decide it's time to return to the automobile. But which way must you go? You really haven't been paying attention to where you walked and did not bother to write the coordinates for the location of your parked car.

There is no problem. You can simply ask your GPS receiver to tell you where you are located and in what direction and how far you must go in returning to your car's location at the stored waypoint you named "CAR." In fact, the unit can tell you the distances and either the map or compass directions among any two of up to 100 or more locations that you may have saved or wish to program into the unit at any time. It can even tell you whether travel between them would be up or downhill by checking the elevations reported for each position's description. Incidentally, you can determine compass directions on the ground (by trial and estimate) using only your GPS receiver, but it is certainly quicker and easier to carry and use a magnetic compass for this purpose (Chapter 4).

Unlike movements at sea or in the air, the quickest way to proceed between two points on land most often will not be a straight line. Mike Hagedorn, a New York State Forest Ranger working in the vast six million-acre Adirondack Preserve in upstate New York (one of the world's largest protected wilderness areas), warns that when selecting a route over land, you must consider its "functional distance." He determines the **functional distance** of any potential route segment through a careful assessment of the **time**, **effort**, and **the level of difficulty** required to move over it. For example, there may be a cliff, swamp, or a steep hill located directly between you and the car. This means the selection of a less direct route will most likely be longer but faster and less difficult to negotiate. In the case of military movements, tactical situations further complicate your routing decisions.

In regard to highway trail and travel, we know that few roads follow straight lines. Also, some roads are easier and quicker to drive than others, so you are always confronted with choices and decisions when planning any route.

Obviously, Step 2 (plan the route) is made easier by using the Magellan equipment because it can quickly

MAGNETIC LENSATIC COMPASS
STOCKER AND YALE, INC.
SALEM, NH U.S.A.

tell you the relationship between your location (your **starting point**) and anywhere you wish to end up (your **objective**) in terms of direction and distance. However, it is just as obvious that you must also know how to use a map in order to effectively apply this information to the selection of a good, **practical** route over the ground. You will find Chapter 3 (Navigating With GPS And A Map) quite helpful in this regard.

STEP 3 (ROUTE FOLLOWING)

Without question, the essence of navigation is staying on the route. You will recall that there are three movement techniques generally used for accomplishing this task - terrain association, dead reckoning, and highway/trail following. They may, of course, be used separately or in concert.

As long as visibility conditions are good (usually related to light and weather conditions or the density of vegetation) and adequate concentrations of identifiable terrain and other landmark features are available in the immediate area, experienced navigators prefer to move by **terrain association**. When employing GPS, this will still be the case, but with GPS, all the uncertainty is gone. You can get a reliable position-fix at any time. The terrain will still provide the guidance needed to wend your way over the selected route to the destination, but you can now move more quickly, with greater confidence, and with far less mental concentration on the navigation task itself. You are free to concentrate on other factors such as enjoying the beautiful scenery or getting the job done.

On the other hand, when movement by terrain association is not possible, navigators proceed by **dead reckoning**. Prior to the advent of GPS, they moved by sighting with their compasses on a series of steering marks encountered along the designated azimuth while measuring or pacing-off the prescribed distances over each segment of the route. These steering marks were uniquely-shaped bushes, trees, rocks - anything they could use to guide them along the way. When steering marks could not be seen due to visibility conditions or in barren landscapes, a man was often sent out ahead to create a steering mark along the prescribed azimuth. (NOTE: An **azimuth** is defined as a 0° to 360° angular directional value measured - by a compass out on the ground or by a protractor on a map - clockwise from a north reference line (e.g., north = 0° or 360°, east = 90°, south = 180°, and west = 270°)).

Generally, when using GPS, the compass should be used to set your general direction of progress without regard to specific steering marks because the GPS unit will keep track of your actual progress. However, **when great precision is required** in moving along a narrow corridor, it may be still easier to sight on steering marks with your compass and utilize the GPS receiver as a backup to insure that you do not wander out beyond your safety limits.

If you are following a dead-reckoned course cross-country with your GPS receiver, it will tell you anywhere along the way what directional azimuth to "steer" and how far it is to the next waypoint (checkpoint).

The navigational information this GPS unit will report when a route has been set in its memory include: (1) the name of the leg destination waypoint, (2) a graphic schematic showing north or south, movement direction, and the relative location of the leg destination waypoint , (3) a bearing and distance from the present position (PP) to the leg destination waypoint, (4) TTG (time to go), (5) ETA (estimated time of arrival), (6) VMG (velocity made good - velocity running a line parallel to

PHOTO: U.S. GEOLOGICAL SURVEY

the course line), (7) SOG (speed over ground - speed of movement in relation to the earth's surface, and (8) the amount and direction of your deviation from the planned course.

Before GPS was available, the navigator **traveling on paths, streets, and highways** was required to recall, identify, and negotiate key turns by spotting, in time, the various landmarks and signs previously keyed in his mind to each of these decision points. Given the many other details and events demanding his attention and the lack of information included on most published road maps, this was often a difficult task to accomplish.

With GPS, staying on all the correct pathways or roads is a simple matter of periodically checking your position using the receiver unit and mentally placing yourself on the map. You'll know well in advance when you are approaching a key turn, route change, or the destination. If you do happen to miss a turn or drive by a desired stop, you can quickly determine where you are located, check the map, select a return route, and get back to the plan with little difficulty.

STEP 4 (DESTINATION RECOGNITION)

The final step is to recognize the objective. Before the development of GPS, unless you were familiar with your destination, it was frequently no easier to identify than any of the other checkpoints, critical turns, or intermediate objectives you were required to find along the way. Fortunately, now with GPS, recognizing these crucial locations is no longer such a challenge.

As part of its route-following function, the GPS receiver **informs you when you have arrived** at your intermediate and final objectives. More specifically, the unit tells you when you have crossed a line drawn perpendicular to your direction of travel through the waypoint located at the end of each route segment. It reports "ARRIVED" and then automatically begins to navigate the next leg of the route and render its reports regarding your progress as soon as the previous one has been negotiated.

THE STEADFAST LN RULE

Always navigate with a correctly oriented map. There is no reason to perform the difficult mental gymnastics necessary to get properly oriented on a map that does not display its features in the same pattern as those it represents in the real world. You may have already attempted this frustrating exercise while examining an improperly oriented graphic floor plan shown on the directory of a large airport terminal or shopping mall.

Even when employing GPS, it is important to keep your map properly oriented. Although you can get a position-fix at any time and may do so frequently, you will find that you should still make good use of terrain and other features, your map, and magnetic compass bearings to guide your movements between these position checks. It is not practical to continuously operate and read the GPS equipment when other matters may require your attention. Use of GPS will give you more freedom from concentration on the navigational task - not less.

You will learn specifically how to orient your map in Chapter 4 when we discuss the use of the compass.

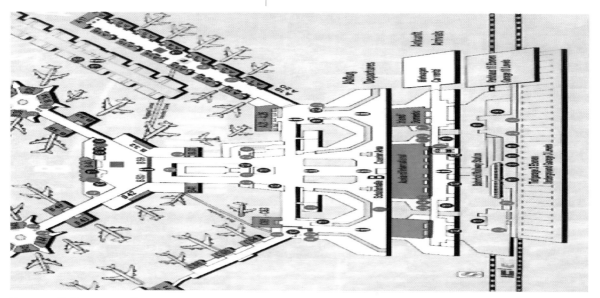

TERMINAL B, FRANKFURT, GERMANY
LUFTHANSA BORDBUCH

ADDITIONAL
FEATURES, CHARACTERISTICS AND OPTIONAL ACCESSORIES OF GPS UNITS

GPS receivers have many valuable features and characteristics that enhance their utility as tools for the land-based navigator. First, they can provide position updates every second and their average position accuracies are within about 25 meters (disregarding the U.S. Government's Selective Availability (SA) defense security policy). Next, they report on the strength of the signals received and the quality of the geometry used (based upon the positions of the satellites) to calculate this data. These units will warn you when a fix should not be used or when signal strength is so weak that contact with a particular satellite may be lost. It should be noted, however, that signal strength has little impact upon the accuracy of a unit's calculation of a position.

Furthermore, these units will track up to 12 satellites, allow you to either automatically or manually store an inventory of literally hundreds of waypoints (landmarks), and enable you to backtrack through the dozen or so positions held in a "lastfix buffer." In addition, these units permit you to establish and navigate multi-leg routes using waypoints you have held in memory. Their special navigational features include on- screen graphic steering aids, course correction information related to direction and distance, a "go to" feature that acts as an instant single leg route from your present position (PP) to any stored waypoint, and various time/speed/distance reports.

Finally, today's portable GPS units fit into the palm of your hand as well as being lightweight, rugged, and waterproof (non-submersible). Most models have large keys for use with gloves and backlit LCD graphics display screens with contrast and light intensity controls. They also come with lanyard straps and carrying cases and offer a variety of accessories such as external power supplies and antennas.

Most GPS receivers are incredibly easy to use and will revolutionize the way you approach your outdoor adventures. Now you can get where you want to go—anywhere in the world, at any time, and under any conditions—simply, quickly, and with absolute confidence. Whether you are deep in the wilderness or moving along a trail or highway, it will give you a quick, accurate fix on your position and other vital information to help you proceed along the way. Magellan's GPS 2000 serves as an excellent example of the newly developed low-cost GPS receiver unit now reaching the market.

IN SUMMARY

If you are an experienced navigator, GPS receivers will be easy to use and integrate into your well-established repertoire of knowledge and skills. It will also eliminate the guesswork, uncertainty, and demand for focused concentration with which you have always had to contend in continually tracking your position and staying on the route. Your navigational quickness, safety, and assurance for success in arriving on time at your destination with a minimum of difficulty will greatly increase. The number of mistakes you make, amount of time wasted while enroute, and stress level related to directing any movement will significantly decrease.

If you are a new or somewhat inexperienced navigator, GPS will make it possible for you to achieve the navigational prowess previously attained by only the most gifted practitioners. This is the case because the GPS unit derives all the vital information formerly available only to individual navigators using every meager clue they could glean from years of study and practice in using the terrain, map, and compass. Simply applying this vital information now provided by the GPS receiver unit - the **easy** part of LN - is all that is left to be accomplished by you.

ILLUSTRATION BY: MATTHEW DWYER

WORLD MAP FROM THE FIRST COMPLETE PRINTERD ATLAS, THE 1482 ULM EDITION OF COSMOGRAPHIA

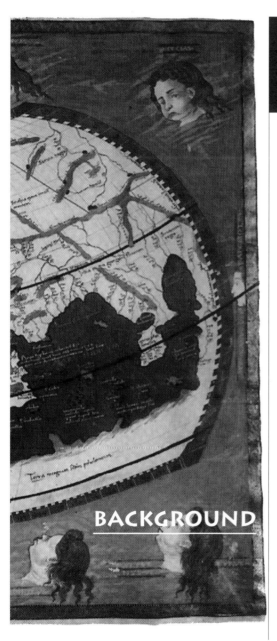

BACKGROUND

NAVIGATING WITH GPS AND A MAP

This chapter will help you to read and interpret maps so you can effectively navigate on land while using GPS. Its objectives are to teach you to (1) locate and report your position using map coordinates, (2) understand the language and techniques used by map makers to portray real world features and convey other related information, and (3) apply a carefully researched strategy for identifying specific terrain features - both when encountered on topographic maps and on the ground - as guides to your movements.

Psychologists remind us that we must always deal with two forms of reality - the one we think exists and the one that is actually out there. When our mind's reality deviates too much from the genuine one, we

31

cease to function effectively at whatever we are doing. That is especially true for navigation. The conceptualizations we form in our minds about the physical shape and content of the real world area in which we are operating are the only basis we have for our navigational decisions and actions.

In order to navigate effectively in any area, it is a great advantage to know it as well as your own neighborhood. Other than living somewhere for an extended period of time, the only way you can seize this "home court" advantage is through an ability to use maps.

Further, without the information maps convey, GPS position reports are nothing more than a meaningless series of letters and numbers. Also, GPS cannot tell us where we may wish to go or which route is the most functional to take us there. We must decide these things for ourselves after consulting a map. Although the GPS receiver reports our location, it is the map that tells us if we are located where we wish to be. And, if we are not, it allows us to determine how best to proceed in making our corrections.

Unquestionably, land navigators using GPS need maps just as medical doctors utilizing advanced CAT SCAN technology still require a comprehensive understanding of the human body. Without them, the vital information being provided by the sophisticated tool has little meaning.

FINDING & REPORTING POSITION

All maps use colors, symbols, labels, and marginal notes to portray the real world around us. But **good** maps also include a coordinate system that facilitates our ability to locate and report any position within the areas covered. In fact, when employing GPS, this becomes the map's most fundamental use. The coordinate systems used for this purpose must be understood by you, any other persons to whom you may wish to communicate your location, and the GPS receiver you employ.

The idea is that the coordinate system used on the map allows you to apply the locational information developed by your GPS equipment to the matrix of lines placed there to define it. In turn, the map's portrayal of the real world's features, as drawn and labeled within that network of lines, allows you to further relate that position to the many characteristics found on the ground in the area through which you are navigating. These characteristics may include both natural and cultural features such as hills, valleys, ridges, bodies of water, forests, trails, highways, towns and cities, man-made structures, and so forth.

There have been several coordinate systems developed over time to define the position of any point on the globe or map. For example, it is quicker and easier to tell someone that you will meet them at the intersection of 2nd Avenue and 8th Street, rather than saying they will find you in front of the 12th white house north of the museum (**Figure 3-2**). Urban street patterns form a simple, yet usable, grid coordinate system. Their limitations are that they are irregular in street spacing and can not be used with a worldwide electronic navigational system or even a small scale map that does not show all the streets and their names. Nevertheless, the concept behind the use of this street grid is very similar to that of reading the standard coordinate systems used by cartographers (map makers) and by GPS units.

A large percentage of the world's base information maps (including most large and intermediate-scale topographic maps) display both geographic coordinates (latitude and longitude) and the Universal Transverse Mercator Grid (UTM) or some similar regional coordinate system (e.g., the British and Irish grids). For example, the U.S. Geological Survey (USGS), as do many other nations' mapping programs, includes the UTM Grid on all published maps at a scale of 1:1,000,000 and larger (**Figure 3-3**).

These maps are frequently used by those engaged in hiking and other forms of outdoor recreation, park and forest management, and search and rescue operations. Just as importantly, they are also used by commercial map producers in their development of standard highway, street, and other types of travel maps. Thus, either geographic or UTM grid coordinate information is easy to

FIGURE 3-2

SEGMENT OF LE MARS, IOWA
1:24,000 SCALE
USGS.

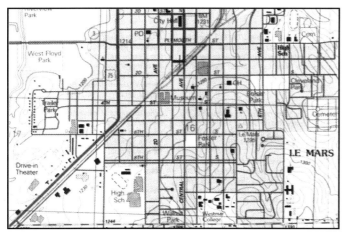

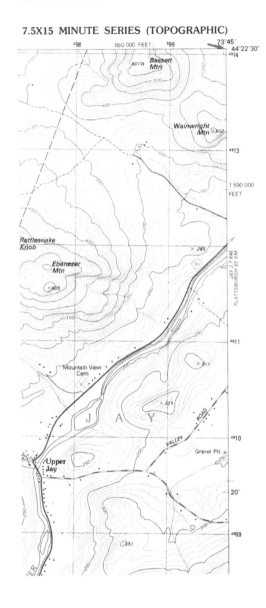

FIGURE 3-3 SEGMENT OF LAKE PLACID, NY
1:25,000 SCALE
USGS

obtain and add to your maps, should it not already be found there. No doubt, commercial producers will soon add this information to their maps as the increased employment of GPS creates a demand for its inclusion.

In the military arena, nearly all large-scale topographic maps around the world include both geographic coordinates and the UTM or some similar regional grid in conjunction with the associated Military Grid Reference System (MGRS) information. Please note that the MGRS is actually a simplified variation of the UTM grid, or some other locally employed coordinate system, that makes it much easier to use. This will all be clearly explained over the next few pages.

Magellan's Trailblazer™ series and GPS 2000 units are ideally suited for use by land navigators because it is designed to report positions in the two most commonly used coordinate systems found on maps: (1) Latitude and Longitude (Geographic Coordinates) and (2) Universal Transverse Mercator Grid (UTM). One of the models in the Trailblazer™ series, the Magellan M5, and one of the Trimble Scout units report UTM grid coordinates in the Military Grid Reference System (MGRS) format as well.

Next, you will learn to read and use these coordinate systems.

GEOGRAPHIC COORDINATES

Reporting positions in terms of their **latitude** and **longitude** has been the fundamental method for defining a point on the earth's surface since it was developed by the ancient Greeks. The positional values for this coordinate system are defined within the context of the network of lines formed by: (1) **Parallels** - horizontal lines drawn east and west around the globe and equidistant from each other (**Figure 3-4**), and (2) **Meridians** - vertical lines drawn north and south on the globe perpendicular to the parallels and converging at the poles (**Figure 3-5**).

Figure 3-6 puts them together to form a complete geographic coordinate framework. The **latitude** of a point on the earth's surface is its distance north or south of the **equator**, as measured by the horizontal parallels encircling the globe. You simply count the values of the lines found between the equator (0° parallel) and the point you wish to define either in the northern or southern hemisphere. (For example, the latitude of Mainz, Germany, is 50° N. Lat. and of Montevideo, Uruguay, is 35° S. Lat.)

On the other hand, the **longitude** of a point on the earth's surface is its distance east or west of the **prime meridian** (0° meridian), as measured by counting the values of the several vertical meridians running between the poles from the prime meridian to the point being considered either in the eastern or western hemi-

FIGURE 3-4

FIGURE 3-6

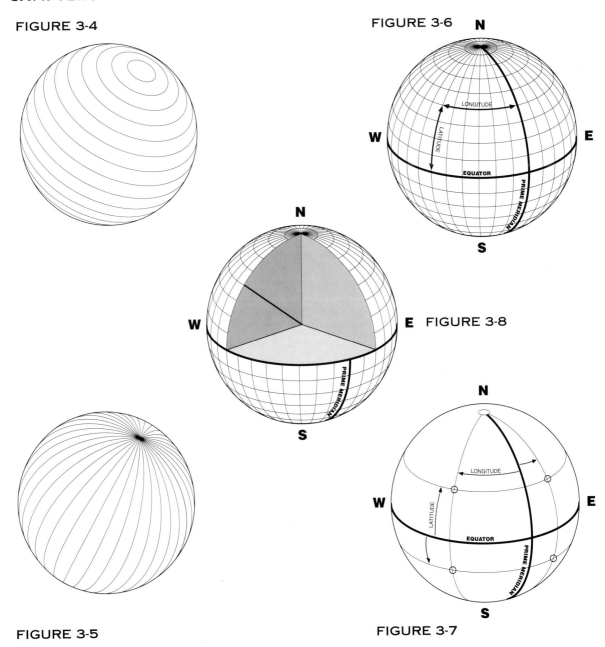

FIGURE 3-8

FIGURE 3-5

FIGURE 3-7

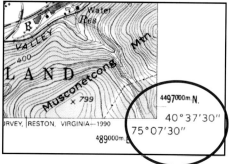

FIGURE 3-9

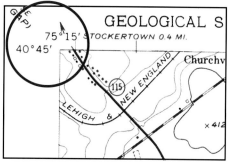

FIGURE 3-10

sphere. Most nations recognize the meridian passing through Greenwich, England, as their **prime meridian.** (For example, the longitude of Aaman, Jordan, is 36° E. Long. and of Milwaukee, WI, USA, is 88° W. Long.)

Figure 3-7 clearly illustrates that, when using geographic coordinates, locations anywhere on the earth's surface are reported as either **north** or **south** latitude and **east** or **west** longitude. **Figure 3-8** gives us a better look at this coordinate framework and conveys the angular nature of the geographic coordinate system.

As illustrated in **Figure 3-8**, latitude is measured in degrees north or south of the equator as if you were to read the angular measurement to any point on the earth's surface from its center. The equator would be level with your horizontally outstretched arm and considered to be 0° (0 degrees) latitude, while the north pole would be vertically straight up at a 90° angle. Thus, the north pole would have a latitude of 90° north. Of course, the same would be true south of the equator where latitude has values ranging from 0° to 90° south. Using this same concept of measuring angles to locations horizontally around the earth's surface from a perspective in the center, longitude ranges from 0° at the prime meridian (Greenwich, England) to 180° both east and west—depending in which hemisphere you are located. **The International Dateline** is located at 180° of longitude. (This imaginary line establishes the beginning of a new calendar day.)

It's easy to see that every point on the globe can be defined and reported in terms of north or south latitude and east or west longitude. However, when using only full degrees of arc to define a position, this system is not very precise. One degree (1) of latitude anywhere on the earth's surface and one degree (1) of longitude at the equator are equal to a distance of 69.17 miles (111.32 kilometers).

If we are going to use latitude and longitude as a means for locating or reporting our locations precisely on a map, we must divide these measurements into smaller units. Therefore, each degree of latitude and longitude is broken into 60' (60 minutes) and each minute into 60" (60 seconds). This has nothing to do with time—they are units of angular measurement. Therefore, 1' (1 minute) of latitude anywhere on the earth's surface and 1' (1 minute) of longitude at the equator are both equal to 1.15 miles (1.86 kilometers) of linear distance. And finally, 1" (1 second) of latitude anywhere on earth's surface and 1" (1 second) of longitude at the equator are equal to a linear distance of 33.82 yards (30.33 meters). Most GPS units can report the latitude and longitude of positions in degrees, minutes, and seconds or in degrees and minutes (down to hundredths of a minute). **Note to Nautical Navigators: 1 minute of latitude anywhere or longitude at the equator equals 1 nautical mile along the earth's surface.**

On most large-scale civilian and military topographic maps, the latitude and longitude of the maps' four corners appear in the margins. For example, the southeast (lower right) corner of the Easton, PA, NJ,

USA, 1:24,000-scale U.S. Geological Survey (USGS) topographic map sheet is located at 40°37'30" N Lat. and 75°07'30" W Long (**Figure 3-9**). The entire sheet covers an area equal to 7.5' of Lat. x 7.5' of Long.; therefore, the northwest (upper left) corner of the map is located at 40°45' N Lat. and 75°15' W Long (**Figure 3-10**). Graticule marks (a network of lines and ticks representing latitude & longitude) are also shown and labeled at prescribed intervals in many maps' margins (**Figure 3-11**).

Using **Figure 3-12**, what is the latitude and longitude of the dot representing the location of that old ship (to the nearest degree)?

Your answer should be 48° N. Lat. and 27° W. Long. Now, using the large-scale topographic map found in **Figure 3-13**, determine the latitude and longitude of the black circle to the nearest minute .

FIGURE 3-11

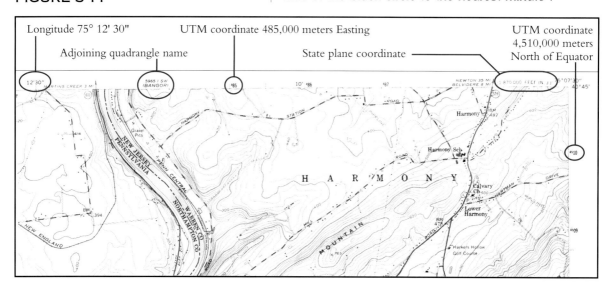

Your answer should be 40° 40' 00" N Lat. and 75° 10' 00" W. Long. Unless you wish to construct or purchase a properly scaled measuring device, you can simply estimate the distances to various locations falling between the parallels and meridians shown on your map or chart.

One disadvantage of using geographic coordinates in land navigation is that 1 degree, minute, or second of **longitude** does not represent the same linear distance along the earth's surface in all locations. They become increasingly shorter as your position moves farther from the equator (both north or south). Remember, the meridians used to measure longitude converge at the poles. For example, 1 degree of longitude equals approximately 69 miles or 111 kilometers **only** at the

FIGURE 3-12

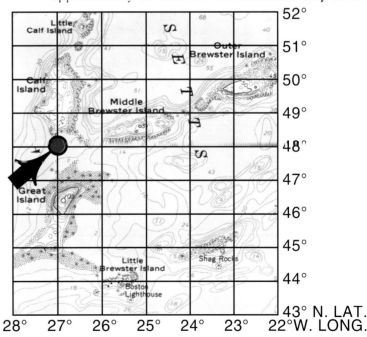

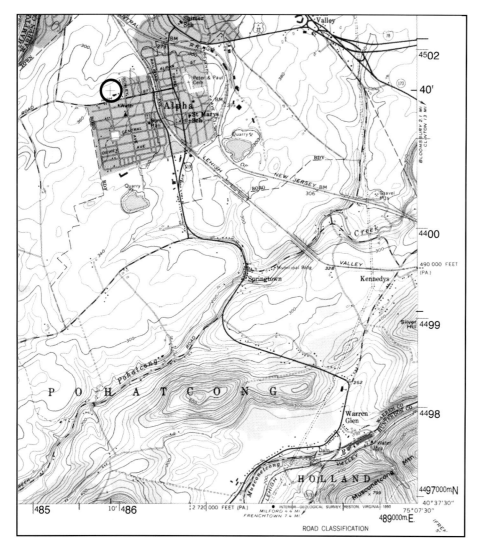

FIGURE 3-13

SEGMENT OF EASTON, PA, NJ
1:24,0000 - SCALE
USGS

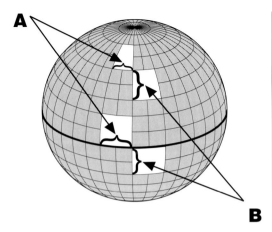

FIGURE 3-14

LONGITUDINAL SECTIONS (A) DO NOT SWEEP OUT
EQUAL DISTANCES AT DIFFERENT LOCATIONS ON THE
EARTH, WHILE LATITUDINAL SECTIONS (B) DO.

UNIVERSAL TRANSVERSE MERCATOR (UTM) GRID COORDINATES

equator. At 85° north latitude, 1° of longitude represents approximately 6 miles or 9.7 kilometers of linear distance (see **Figure 3-14**). On the other hand, linear distances for differences in latitude remain unchanged anywhere.

Now that you fully understand how to locate any position on a map displaying geographic coordinates, we will proceed to examine a coordinate system that was specifically designed for use on land.

Cartographers must always introduce distortion errors into their maps because they are representing the surface of our spherically-shaped planet on a flat piece of paper. Thus, the decision facing any map maker becomes how much distortion error the maps have and of what type. Over the centuries, a multitude of mathematical schemes, called map projections and coordinate systems, have been developed. The coordinate system used most commonly on medium- and large-scale maps produced by the U.S. and many other

FIGURE 3-15

governments is the **Universal Transverse Mercator Grid** (UTM). The Transverse Mercator Projection induces the least amount of distortion on a series of large-scale maps covering sizable land areas and the UTM Grid Coordinate System provides a perpendicular grid with constant linear surface distance values between each of its grid lines in all directions.

To understand how this projection works, imagine the earth as an orange with parallels and meridians drawn upon it. Now, using a knife and after slicing off small circles at the poles, make a series of straight north-south cuts in the peel at equal intervals of 6° completely around the orange until 60 identical strips have been detached (**Figure 3-15**).

Each of these segments forms the basis of a separate map projection. Because each zone is relatively narrow (only 6° of longitude), its flattening results in a minimal distortion of the features shown on the surface.

The UTM Projection has been designed to cover that portion of the earth's surface located between the latitudes of 80° south to 84° north in a wide band running around the globe. This includes most of the world's inhabited lands. There is another grid system designed for the circular areas cut out of the two polar regions (Universal Polar Stereographic Grid—UPS).

By international usage, all of the UTM grid zones have been consecutively numbered from west to east (left to right) 1 to 60, beginning at the **International Dateline** (180° Long.). **UTM grid zone 1** is a vertical area running between the meridians located at 180° W and 174° W Long. (6 degrees in width), with its central meridian being located at 177° W Long. The Easton, PA, NJ, USA, map sheet discussed earlier is located within UTM grid zone 18 (**Figure 3-16**).

Since the pattern of UTM grid lines were superimposed on the "orange peel strips" (grid zones) **after** they were flattened, these grid lines are straight, undistorted, and perpendicular. However, all meridians and parallels, with the exception of the central meridian and the equator (central parallel) within each of these grid zones, were slightly distorted by the flattening process (**Figure 3-16**). This means that only the **central meridian** and **equator** can serve concurrently as lines of latitude and longitude and as perpendicular UTM grid lines. Therefore, these two lines have become the basis for numbering the grid lines within each of the 60 grid zones. In addition, they obviously serve to link the UTM Grid Coordinate System to the True Geographic Coordinate System.

All grid zones' central meridians (177°, 171°, 165°, and so forth) are arbitrarily labeled 500000mE (500,000 meters) to create a west to east numbering system within the grid zone. The west to east (left to right) grid line labels never reach the zero point as you proceed·to the west (left)—they just have decreasing values until you reach the western boundary of the zone. Con-

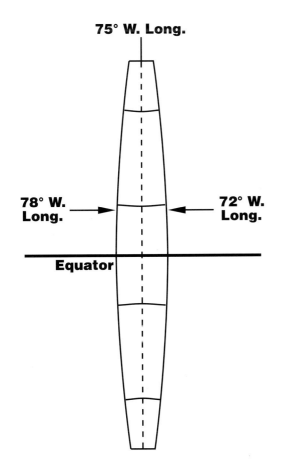

75° W. Long.

78° W. Long. → ← **72° W. Long.**

Equator

FIGURE 3-16

UTM GRID ZONE 18

versely, as you proceed to the east (right) of the central meridian, the grid line label values increase from 500,000 meters until you reach the eastern boundary.

On the other hand, the equator serves as the 0000000mN (0 meters) grid line for the northern hemisphere or 10000000mN (10,000,000 meters) grid line for the southern hemisphere to create a south to north numbering system within each hemisphere of the zone. Remember, grid lines have increasing values as you proceed to the east and to the north (**Figure 3-17**).

Coordinates reported in degrees, minutes, and seconds of latitude and longitude are often called **true** coordinates. Therefore, those reported in terms of the UTM grid are often called **false** (slightly inaccurate) **eastings** and **northings** because of the slight distortions caused by flattening each of the sixty grid zones (orange peel slices) prior to placing the perpendicular UTM grid lines on them.

To briefly summarize what you have just learned about the UTM Grid Coordinate System, each of the UTM grid zones (orange peel slices) is 6° wide. The central meridian and the equator serve as the origin for labeling the false easting and false northing values represented by the grid lines superimposed on each of the 60 UTM grid zones encircling the globe. Each central meridian is labeled 500,000 meters E. (500 kilometers). The equator serves as 0 meters/kilometers N. for the northern hemisphere and 10,000,000 meters/10,000 kilometers N. for the southern hemisphere. Grid values increase as you proceed from west to east and from

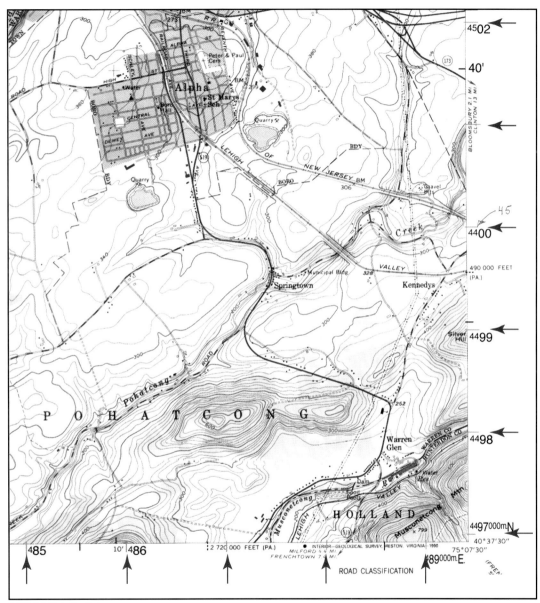

FIGURE 3-17   SEGMENT OF EASTON, PA, NJ
1:24,000 SCALE
USGS

south to north. Therefore, you read UTM grid coordinates to the **right and up**.

Use of the UTM grid coordinate system is much easier for the land navigator to adapt to than is latitude and longitude. The grid is made up of perpendicular lines with equal ground distances from one to the next. They are also clearly labeled on most large-scale and many intermediate-scale maps.

For example, take another look at the segment of the southeast (lower right) corner of the Easton, PA, NJ, USA 1:24,000-scale USGS topographic map sheet shown in **Figure 3-17**. The UTM grid system is represented by small blue tick marks and some black numerical labels in the margins outside the frame (neatline) of the map. (Note the arrows in the illustration pointing to these tick marks.). These tick marks can easily be connected using a pencil and straight edge to form the UTM grid pattern on the map as in **Figure 3-3**. The newer large scale (1:25,000 or 1:50,000-scale) and many of the smaller scale USGS map sheets (but larger than a 1:1,000,000-scale) carry a full printed UTM grid. All large- and intermediate-scale military topographic maps produced by the U.S. Department of Defense Mapping Agency (DMA) have full UTM grid lines printed on them, as well.

Please note that the UTM grid line closest to the east (right) edge of the Easton, PA, NJ, USA, map (**Figure 3-17**) is labeled 489000mE (489,000 meters of false easting). Since the central meridian of each grid zone is arbitrarily labeled 500,000mE, we know that this grid line is only 11,000 meters (11 kilometers) from the central meridian. Because its value is less than 500,000mE, we also know that it lies 11,000 meters (11 kilometers) west of (short of) the central meridian. You will recall that UTM grid values increase from west to east. On the other hand, the UTM grid line closest to the bottom of the map sheet is named 4497000mN (4,497,000 meters of false northing). For the areas covered in the northern hemisphere within each grid zone, the equator is the 00 starting point. Thus, this grid line is 4,497,000 meters (4,497 kilometers) north of the equator.

Why do land navigators find the UTM grid relatively easy to read on the map? Because the **two large numerals** in the grid line labels are generally the only important ones to be concerned with as you navigate within a local area. They represent the informal first names of the grid lines commonly printed 1000 meters (1 kilometer) apart on large-scale civilian and military topographic maps.

For both false easting and northing values, **the two large numerals** that serve as the labels for these grid lines start at 00, run through 99, and then start over again at 00. This means that if you know in which larger 100,000 meter (100-kilometer) square area you are located (or if you don't care because you don't plan to stray more than 100 kilometers away), you can completely ignore those first one or two smaller numerals. If you are concerned about this larger area, you need only pay attention to the **first set of one or two smaller numerals** in order to learn in which 100,000 meter (100 kilometer) square block you are located west to east and south to north within each numbered UTM grid zone. You will then be able to compare the full label UTM grid values reported on your GPS receiver with those on your map in order to confirm your position within a larger area. Regardless, it is still the two large numerals that will tell you your precise location on the map.

The last three smaller numerals (not always included in the label) represent single meter values (0 to 999), but as a land navigator, you will not be interested in grid coordinate readings more refined than 100 meters. For example, the school just north of Warren Glen on the Easton, PA, NJ, USA, map sheet is located in grid square 8898 (**Figure 3-17**). To report that position more precisely to within 100 meters, you would say that it's located at coordinates 488500mE and 4498200mN. Most people just short hand it as 885E 982N. That means it is located half way between grid lines 88 and 89 going east (right) and two tenths of the distance between grid lines 98 and 99 going north (up).

Remember, the rule for reading UTM grid coordinates – **read to the RIGHT and UP**.

MILITARY GRID REFERENCE SYSTEM (MGRS)

FIGURE 3-18

THE MAGELLAN GPS NAV 1000M5

As was stated earlier, although all GPS units do not at this time report positions in the MGRS/UTM grid format, several GPS units do offer this option. In effect, MGRS is simply the UTM Grid System modified and shortened into a simple alphanumeric system that is quicker and easier to understand and use. You can apply the MGRS format to any map displaying UTM grid values, and you can think in terms of MGRS—even when your GPS receiver reports positions in pure UTM grid values. For these reasons, a detailed discussion of the MGRS has been included in this guide.

This "shorthand" approach is exactly why using the Military Grid Reference System (MGRS) on the Magellan M5 is quicker and easier (**Figure 3-18**). For most land navigators and for most LN applications, **MGRS may become the primary coordinate system of choice for land navigators** (civilian and military) using GPS. The reason is that it makes the application of positional readings from any GPS receiver to the map and vice versa a virtual "snap."

Virtually everything you have learned about the UTM grid applies to MGRS. The general concept is that MGRS divides the world into large geographic areas, each of which is given a unique alphanumeric label called the **Grid Zone Designation**. Each of these Grid

Zone Designations, in turn, is covered by a pattern of 100,000-meter (100 kilometer x 100 kilometer) squares—each being labeled by two letters called the **100,000-Meter Square Identification**. Finally, each 100,000-Meter Square Identification is further subdivided by the regular UTM grid lines with which you are already familiar.

More specifically, the sixty 6° wide UTM grid zones (orange peel strips) are further divided in the MGRS from south to north (bottom to top) into 20 lettered horizontal rows, each having a height of 8° of latitude (excepting the most northern row, which is 12°). These then are the building blocks of the MGRS called **Grid Zone Designations**. In **Figure 3-19**, you see that the Evans Mills (Fort

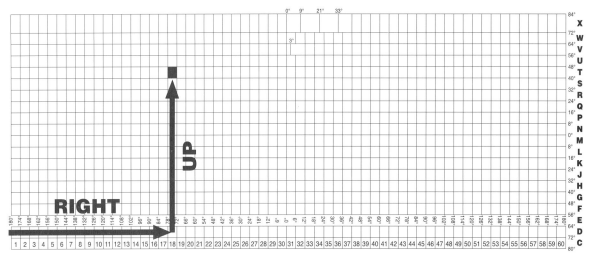

FIGURE 3-19

Drum), NY, USA, DMA map sheet is located in the area covered by Grid Zone Designation **18T** (west to east - right to UTM grid zone 18 and then south to north - up to row T), as is the Easton, PA, NJ, USA, USGS map sheet.

In MGRS, (as with any UTM-based grid) you must read right first and then up. Figure 3-20 shows a number of the 60 vertical grid zones and the 20 horizontal rows as they appear superimposed on a globe. This particular illustration shows us that the Berlin, Germany, map sheet falls within the area covered by **MGRS Grid Zone Designation 33U** (west to east to UTM grid zone 33 and then south to north to row U).

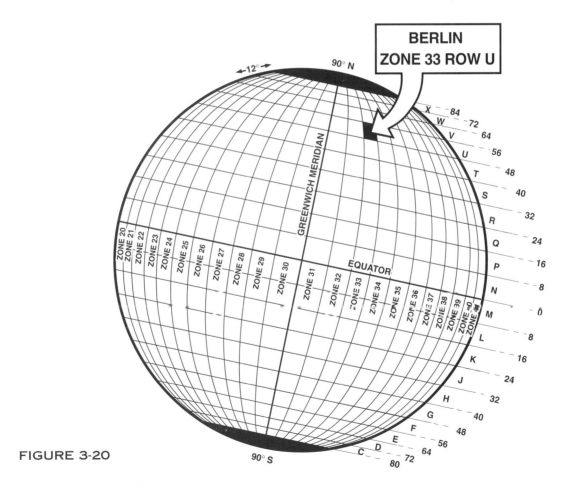

FIGURE 3-20

The basic MGRS pattern of perpendicular grid lines are placed 100,000 meters (100 kilometers) apart and superimposed on each of the 6 degrees of longitude by 8 degrees of latitude (6° x 8°) Grid Zone Designations. This pattern of lines breaks each of these geographic areas into 100,000 meter x 100,000 meter (100 kilometer x 100 kilometer) squares identified by two letters. **Figure 3-21** shows us Grid Zone Designation 18T, the one in which Evans Mills (Fort Drum), NY, USA, and Easton, PA, NJ, USA, are located, subdivided into several full and some pleat-shaped dual-letter labeled **100,000-Meter Square Identifications**. These partial area identifications are created along the pleats of the 6° wide UTM grid zones because the linear widths of the grid zones (orange peel strips) decrease as you proceed north or south from the equator. The illustration highlights the fact that the Evans Mills (Fort Drum), NY, USA, map sheet is located in the 100,000-Meter Square Identification labeled VD.

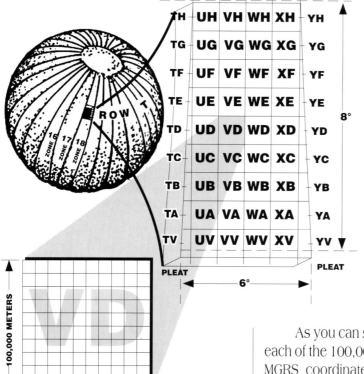

FIGURE 3-21

As you can see, these dual-letter labels identifying each of the 100,000-meter squares contained within the MGRS coordinate system replace those one or two small numerals found in front of the two large numerals in those UTM grid labels you have already studied on your map. When using MGRS, these small numerals can be completely ignored. Now you are concerned **only** with the large numerals on the map's grid line labels. Once you pass beyond a grid line labeled **99** to one labled **00**, when headed either east or north, or fall below a grid line labeled **00**, when headed west or south, you have just passed into the area of another dual-lettered 100,000-Meter Square Identification.

Finally, these **100,000-Meter Square Identifications** are further subdivided by the regular numbered UTM grid lines spaced 1000 meters (1 kilometer) apart on most large-scale topographic maps. Remember, the MGRS requires **only** that you read the large numerals (**00** through **99**) on the UTM grid line labels. When the

FIGURE 3-22

Suez Canal, Egypt
1:250,000 - Scale

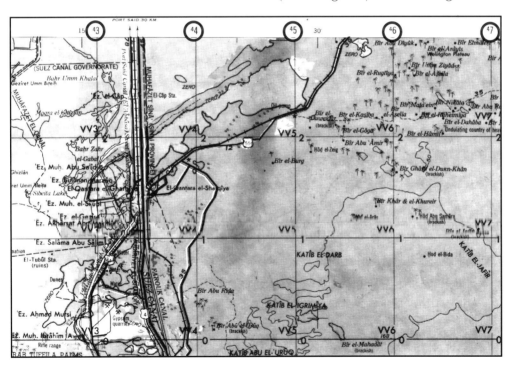

map scale is smaller than 1:100,000, these grid lines **may** be placed every 10 kilometers and numbered with a single large-size numeral **0** through **9** (**Figure 3-22**). When a grid line is labeled by a single large-size numeral, such as **4**, it has the same value (and is the same grid line) as when it is labeled on a larger scale map of the same area as **40**. In other words, grid lines labled with a single large numeral are spaced every 10 kilometers while grid lines labled with two digit large numerals have a spacing of one kilometer

You will note that maps using the MGRS still show the full UTM grid line values (both the small and large print numerals). But, as was stated earlier, **when using MGRS, the small print numerals can be completely ignored**.

GENERALLY SPEAKING, WHERE AM I?

In order to help you easily identify the **Grid Zone Designation** and **100,000-Meter Square Identification** in which you may be located, we have developed a special **MGRS Reference Map** (Appendix A) covering the contiguous 48 U.S. states and areas of southern Canada. The base map shows state as well as county boundaries with overprinted information relating to Grid Zone Designations and 100,000-Meter Square Identifications.

The numbers running from west to east (left to right) across the bottom of the map label the 6° wide **UTM Grid Zones** covering this part of the world. You will

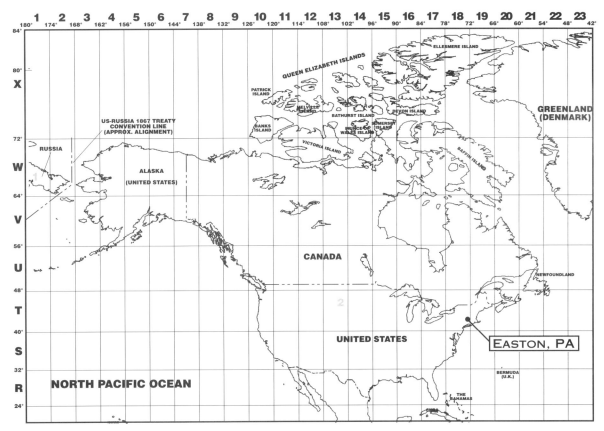

FIGURE 3-23

recall that it takes 60 of these vertical UTM Grid Zones to encircle the entire globe. The letters running from south to north (bottom to top) up the left side of the map label the 8 high rows that complete the definitions of the 6⁵ of longitude by 8⁵ of latitude (6⁵ x 8⁵) **MGRS Grid Zone Designations**. It should be noted that the lines of longitude (every 6⁵) and lines of latitude (every 8⁵) that form the boundaries of these MGRS Grid Zone Designations, as well as the numbers and letters labeling them, are printed on this special map in red.

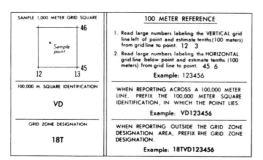

FIGURE 3-24

EVANS MILLS, NY
1:50,000 - SCALE

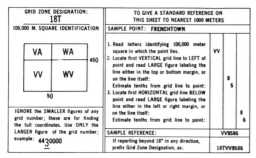

FIGURE 3-25

AL JUMUM, KSA
1:50,000 - SCALE

Within each of these Grid Zone Designations shown on the special map, you can also see the pattern of lines forming and the dual-letter labels identifying each of the **100,000-Meter Square Identifications** contained therein. This information is printed in blue.

A crude representation of this map, displaying Grid Zone Designations and included as **Figure 3-23**, shows us that Easton, PA, USA, is located in MGRS **Grid Zone Designation 18T** (right to UTM grid zone 18 and up to row T). It is also located in **100,000-Meter Square Identification VA.** You will also note that several map sheets covering the Easton, PA, NJ, USA, area are likely to include parts of at least four 100,000-Meter Square Identifications (**VA, WA, VV, and WV**), because their boundaries are very close to the City of Easton. The boundary between 100,000-Meter Square Designations **VA** and **VV** runs approximately 2 kilometers to the south and the boundary between **VA** and **WA** is about 16 kilometers to the east of the City. A typical 1:50,000-scale civilian or military topographic map of any area within the continental United States covers an area of approximately 28 x 22 kilometers (15' Lat. x 15' Long.). Even the very large scale 1:24,000-scale USGS quadrangle (7.5' Lat. x 7.5' Long.) map covering the small 14 x 11+ kilometer Easton, PA, NJ, USA, area includes small portions of two **100,000-Meter Identifications** (**VV** and **VA**) within **Grid Zone Designation 18T** because the area encompasses the boundary of these two MGRS subdivisions.

Maps produced by the U.S. Department of Defense Mapping Agency Topographic Center (DMATC)

FIGURE 3-26

NEWARK, NJ, PA
1:250,000 - SCALE

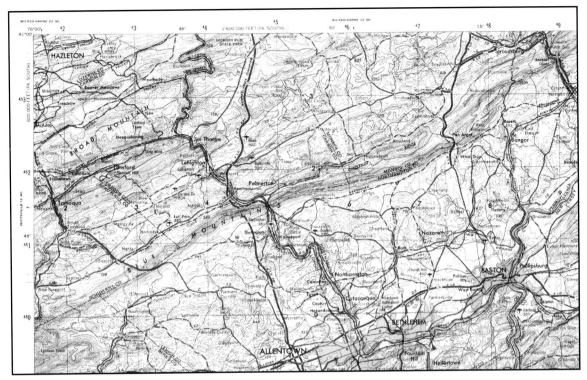

FIGURE 3-27

NEWARK, NJ, PA
1:250,000 - SCALE
USGS

and numerous other mapping agencies around the world include a **grid reference box** in their margins. **Figure 3-24** was taken from the margin of the Evans Mills (Fort Drum), NY, USA, 1:50,000-scale map sheet, **Figure 3-25** from the Al Jumum, Kingdom of Saudi Arabia, 1:50,000-scale map sheet, and **Figure 3-26** from the Newark, NJ, PA, USA, 1:250,000-scale map sheet (U.S.G.S.). The medium-scale Newark map sheet encompasses the Easton, PA, NJ, USA, area used in previous illustrations.

You will note that the area covered by the Evans Mills (Fort Drum) map sheet is located within the Grid Zone Designation **18T** and 100,000-Meter Identification **VD**. The area covered by the Al Jumum map sheet is located within the Grid Zone Designation **37Q** and the 100,000-Meter Identifications **EE** and **ED**. And, the area covered by the Newark 1:250,000-scale map sheet is located within the Grid Zone Designation **18T** and the four 100,000-Meter Square Idetifications: **VA**, **WA**, **VV**, and **WV**. Please note that the grid line labels on the 1:250,000-scale map sheet have only one large numeral because the grid lines are spaced every 10,000 meters (10 kilometers), rather than every 1000 meters (1 kilometer), as on larger scale maps (**Figure 3-27**).

In summary, by examining and recognizing the state and county boundaries found on the useful MGRS Reference Map (Appendix A), you should be able to relate any regional location encompassed by the map to the general framework of the MGRS which has been superimposed and color coded on it.

All maps and GPS receivers in current use that include references to MGRS Grid Zone Designations and 100,000-Meter Square Identifications present the information in three ways. MGRS Option 1 displays position coordinates using the Designations and Identifications used on maps produced before and during 1987. MGRS Option 2 utilizes the labels that were in use on maps produced between January, 1988 and September, 1991. MGRS Option 3 covers those maps produced since October 1991 (currently, there are few). Appendix A includes two MGRS Reference Maps printed on opposing sides of the sheet showing MGRS Options 1 and 2. Therefore, these reference maps identify general position information for all maps produced prior to September, 1991. As significant numbers of maps are produced using Option 3, another reference map will be produced to present this information.

MORE SPECIFICALLY, WHERE AM I?

Now, let's get down to the useful details. When using a large-scale civilian or military topographic map showing a MGRS/UTM grid every 1000 meters (1 kilometer), we begin to pinpoint locations by first locating the grid square in which the position falls. For example, the "X" marking the top of hill 450 on the map segment in **Figure 3-28** is located in the 1,000-meter grid square "named" 1143 (not 4311) because you must **read right first and then up**. The first two digits naming this grid square represent the false easting grid label (read from left to right), and the last two digits represent the false northing label (read from the bottom to top).

To be even more precise in your position reporting, you can subdivide the 1000 meter square in each direction into tenths (100 meter segments), either by measurement or estimate, and specify the position reading in greater detail. Hill 450 is located .6 of the distance between grid lines 11 and 12 (going east) and .3 of the way between grid lines 43 and 44 (going north). Thus, the coordinates of "X" to the nearest hundred meters, by **reading right first and then up**, are reported as 116433 .

If this position were located in New Jersey some distance northeast of Easton, PA, USA, its full designation in the MGRS (option 1) would be 18TWA116433. By including the Grid Zone Designation (**18T**), 100,000-Meter Square Designation (**WA**), and six-digit numerical coordinates (116433), you have given the "X" on hilltop 450 a unique world-wide address to a degree of accuracy of 100 meters. The MGRS readout on the screen of the GPS receiver would be (18TWA115 433). It would also report its elevation as being 450 feet or 137 meters above sea level, depending upon which units are used to report elevations on the map and were set on the receiver. You certainly should have no difficulty reading and understanding this information. When you go to pinpoint your location on the map, you simply use the two (2) large numerals in the grid line labels and interpolate the number of tenths between them for the third and sixth digits of the six-digit coordinate reading.

Incidentally, 116433 is referred to as a six-digit coordinate for obvious reasons. As a land navigator, you are generally required to work with and report positions on a map in six-digit coordinates (to within a 100-meter square area), but MGRS/UTM coordinates can be further refined.

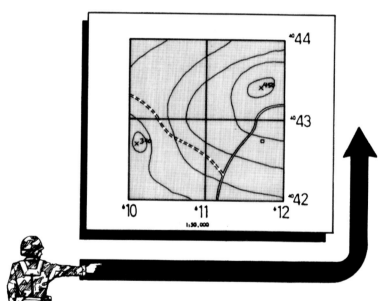

FIGURE 3-28

FIGURE 3-29

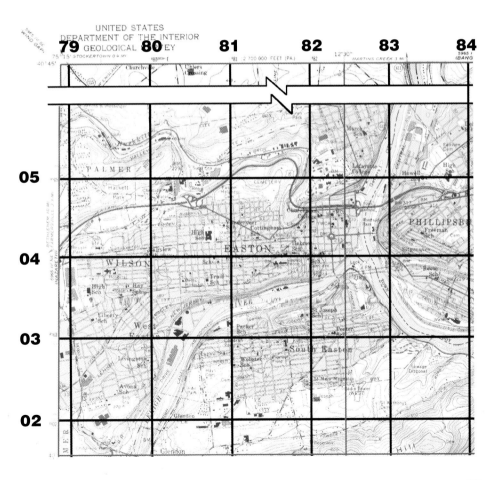

In using the Military Grid Reference System (MGRS), a position is generally written as an entity without spaces, parentheses, dashes, or decimal points. For example:

18T	Locates a point within the 6° x 8° Grid Zone Designation.
18TVA	Locates a point within a 100,000-meter square.
18TVA80	Locates a point within a 10,000-meter square.
18TVA8205	Locates a point within a 1,000-meter square.
18TVA825052	Locates a point within a 100-meter square.
	(If you are a surveyor, rather than a navigator, you might wish to continue.)
18TVA82500527	Locates a point within a 10-meter square.
18TVA8250105270	Locates a point within a 1-meter square.
	(And so forth.)
Note 1:	MGRS grid coordinate 18TVA825052 is the location of the newly constructed Farinon Student Center on the campus of Lafayette College in Easton, PA, USA, (see **Figure 3-29**). Also, please note that its UTM Grid Coordinate is Grid Zone 18 482500E 4505200N. Finally, its true geographic coordinate is approximately 40° 41' 51" N. Lat., 75°12'26" W. Long.
Note 2:	Although MGRS grid coordinates are written without punctuation or spaces, GPS units often report them with a space separating the false easting from the false northing values (e.g., the Farinon Center is located at grid coordinates **18TVA825 052**).

Using the segment of the Lake Marian, FL, USA, map sheet found in **Figure 3-30**, estimate the six-digit coordinates of the features located at points 1 through 4. Remember, to **read right first and then up**. The Grid Zone Designation of this area of Florida is **17R** and the 100,000-Meter Square Identification for the area shown on the map is **MA**.

The solution to the above exercise is as follows: The position of Pt. 1 = 17RMA941871, Pt. 2 = 17RMA957862, Pt. 3 = 17RMA936838, and Pt. 4 =

17RMA965833. When navigating within a local area, the MGRS grid coordinates can be read and reported by using only the 100,000-Meter Square Identification (MA) or only the six final digits of the position's coordinates (e.g., Pt. 4 may be reported as MA965833 or as 965833).

BRIEF RECAP

Always knowing where you are in every possible way (relative to a locational grid, the map, and the real world) is imperative for success in land navigation. You now understand how to locate any position on a map using latitude and longitude (Geographic Coordinates) and UTM or MGRS grid coordinates. Now, when your GPS unit reports your position, you can easily find it on the map and then quickly relate it to the many features portrayed there and actually found on the ground in the area surrounding you. These features, whether they are natural or man-made, can be used to guide and channel your movements over functional routes to your objective or destination.

Now that you are able to use coordinate systems to locate and report your position on a map, you are ready to consider the many other aspects of map using to assist you in navigating over the land.

SELECTING A GOOD MAP

A good map must include as much useful information as possible to help guide your movements without

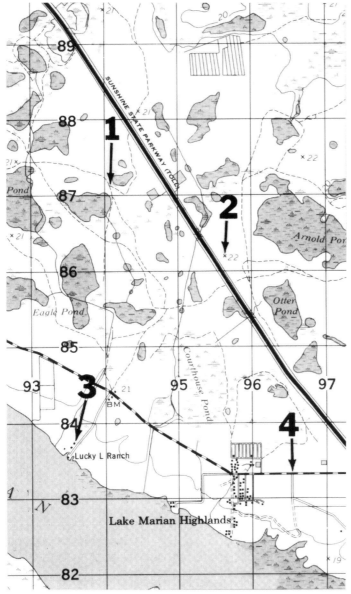

FIGURE 3-30

being cluttered and difficult to read. When selecting a map, you should consider the following: (1) its scale, (2) the amount of detail shown, (3) the quality of its portrayal (accuracy and legibility), (4) the reputation of its publisher, (5) its compilation date, and (6) whether it displays a coordinate system you understand and is compatible with your GPS equipment.

In regard to **map scale**, it should be as large as is practical depending upon the types of movements you are planning to make. You will recall that map scales are reported as a fraction or proportion (e.g., 1/50,000 or 1:50,000). The numerator represents the number of units measured on the map as compared to the number of those same units found in the denominator and out on the ground. Understand that a 1:50,000-scale map is larger in scale than one of 1:100,000-scale (1/2 pie is more than 1/4 pie). It will show more features and in much greater detail, but it will cover less ground area (**Figure 3-31**).

Dismounted navigators prefer to use large-scale topographic maps of 1:25,000- or 1:50,000-scale (**Figures**

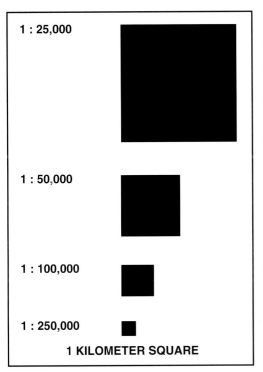

1 : 25,000

1 : 50,000

1 : 100,000

1 : 250,000

1 KILOMETER SQUARE

FIGURE 3-31

3-32 & 3-33), while vehicular-mounted navigators, who are required to cover greater distances, may choose to use a smaller scale map. This might be a 1:75,000- or 1:100,000-scale county topographic or highway map (**Figure 3-34**). Finally, long-distance turnpike or freeway drivers may select a 1:250,000-scale topographic map (**Figure 3-35**) or a 1:1,000,000-scale state highway map (**Figure 3-36**) because a high degree of precision in pinpointing their locations is less important. Utlimately, scale can become so small that the map (or in the case of **Figure 3-37** - the photograph) has little value to the navigator.

Generally, the more **detail of navigational significance** included on your map, the more useful it will become in helping you to select your route and guide your movements. The cross-country navigator must have information relative to terrain, water, vegetation and cultural (man-made) features. Those following trail and road networks can also benefit from this same type of information. However, those moving at a rate of 55 to 65 miles per hour or 90 to 100 kilometers per hour over well-marked thoroughfares can manage with far less detail. Nevertheless, the more information provided by the map, the easier it is to make good routing decisions (in terms of both actual as well as "functional" distances) and to follow that route.

The **quality of a map's portrayal of real world features**, including both legibility and accuracy, is not too difficult to judge by closely examining it. Too much information, inclusion of details not related to LN, and poor choices in the use of colors and symbology all

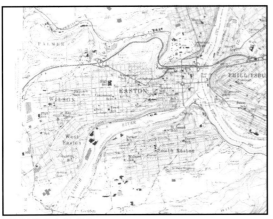

FIGURE 3-32

SEGMENT OF EASTON, PA, NJ
1:24,000 SCALE
USGS

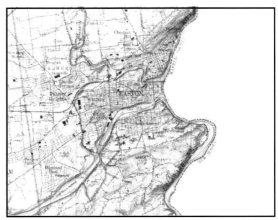

FIGURE 3-33

SEGMENT OF NORTHAMPTON COUNTY, PA
1:50,000 SCALE
USGS

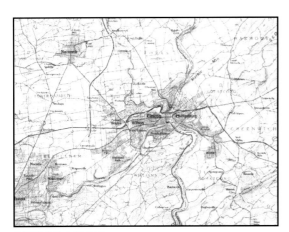

FIGURE 3-34

SEGMENT OF ALLENTOWN, PA
1:100,000 SCALE
USGS

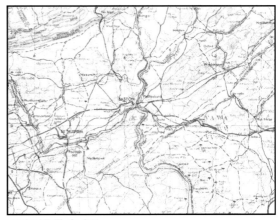

FIGURE 3-35

SEGMENT OF NEWARK, NJ, PA
1:250,000 SCALE
USGS

FIGURE 3-36

USGS

contribute to the degree of difficulty encountered when reading a map. Accuracy is a bit more difficult to judge, but there are also several items you can check quickly to determine a map's quality in this regard.

Whenever the scale of the map is relatively large, look to see if the roads, streams, shorelines, and the edges of any forested areas portrayed seem to show frequent and somewhat irregular ripples and bends. Nature produces few straight lines or sweeping curves. In other words, the appearances of the many details included should be in relatively sharp focus rather than appearing to be highly generalized. Examine the map closely to determine if there is a fair amount of detail shown as compared with the amount of space available in which to portray it. Large open areas on the map are an obvious warning sign regarding its comprehensiveness. Finally, closely examine any portion of the map covering an area with which you are familiar and compare its accuracy against your own knowledge of the area. If the map meets these quick tests with rather high marks, its accuracy can generally be relied upon to guide your movements.

Without question, the **reputation of the map's producer** for high quality cartography work can also serve as a guide to the map's potential accuracy. Today, most maps produced by governmental agencies are highly accurate. Do not trust any map that does not carry a credit line somewhere in its margin. If the producer doesn't want you to know who did the work, you probably should not rely upon it either.

FIGURE 3-37

SEGMENT OF SATELLITE IMAGE MOSAIC
NJ, PA
1:1,700,000 SCALE
USGS

Although a map's **compilation date** is not related to the quality of work done in its preparation, it is, nevertheless, directly related to the accuracy you might expect from the map. This is especially true in regard to its portrayal of cultural features, such as buildings and highways, as well as to any information detailing vegetation. Both culture and vegetation are frequently and significantly changed over time. On the other hand, terrain and water features (hydrography) are only rarely changed in any significant way over lengthy periods of time.

When using GPS, the **inclusion of a compatible coordinate system on the map** is of paramount importance, but there is a reason why it was listed last among those factors to be considered when selecting a map. The reason is that any map user can easily add either the MGRS/UTM grid or lines of latitude and longitude to any map they may prefer to use. The wrong scale, sparse detail, a poor quality of portrayal, and grossly inaccurate or outdated information are problems that are not so easy to rectify.

PREPARING ANY MAP FOR USE WITH GPS

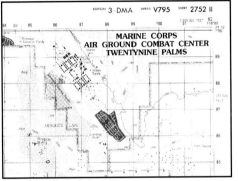

FIGURE 3-38 SEGMENT OF TWENTYNINE PALMS
1:50,000 SCALE
USDMA

FIGURE 3-39 SEGMENT OF ALLENTOWN, PA, NJ
1:100,000 SCALE
USGS

When using government-produced civilian or military topographic maps (particularly those produced by the USGS or DMA), you will frequently have the MGRS/UTM grid lines printed across the face of the map (**Figures 3-38, 3-39, & 3-40**). But when they are represented only by blue tick marks in the margins outside the neatlines of the maps (as on USGS 1:24,000- and 1:62,500-scale quadrangles), you can simply take a straight edge and carefully connect the opposing tick marks to form the familiar grid pattern. You may also wish to label each with larger, more legible numbers using only the two large print numerals found as part of the MGRS/UTM grid labels that are already printed in the map's margins (**Figure 3-41**).

Until such time as the expanding use of GPS encourages commercial map producers to include MGRS/UTM grid lines on their street and highway maps and the various motoring clubs do so on their strip trip maps, you may have to add these lines yourself. It is certainly not difficult to accomplish.

To prepare a localized area street or any other type of map for use with a GPS receiver, refer to large-scale base maps of the area covered by your map. These base maps may be USGS or DMA quadrangle map sheets (**Figure 3-42**). You then add the MGRS/UTM grid lines every kilometer (1000 meters) by "copying"

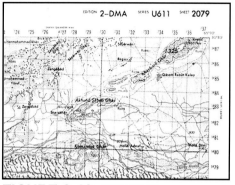

FIGURE 3-40 SEGMENT OF
DO AB, AFGHANISTAN
1:100,000 SCALE
USDMA

the grid lines found on the reference base maps in their relationships to the various features portrayed on both maps. Draw each of these grid lines on your street map as straight lines connecting three or four points you have identified along the path of the matching grid line displayed on the reference map. These reference points must be identified on both maps and the common points carefully located on your new map in relation to selected features shown on both maps. For example, you may use a unique curve along a road, power transmission line, stream course, or shoreline; a high-

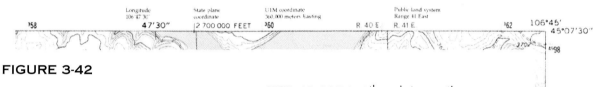

FIGURE 3-42

way or some other intersection; or a manmade structure, such as a building or bridge (**Figure 3-43**). If you wish to use geographic coordinates rather than the MGRS/UTM, lines of latitude and longitude can be added to your map in precisely the same manner.

To prepare highway maps or strip trip maps covering an entire state or region, obtain smaller scale government reference base map sheets, (e.g., USGS/ DMA 1:100,000-, 1:250,000-, or 1:1,000,000-scale maps) and add the UTM/MGRS grid lines every 10 kilometers (10,000 meters), just as they appear on the reference

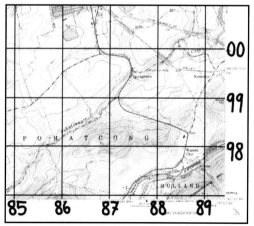

FIGURE 3-41 SEGMENT OF EASTON, PA, NJ
1:24,000 SCALE
USGS

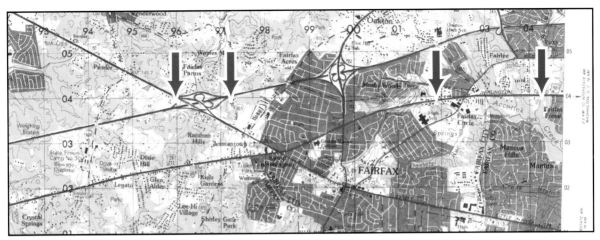

FIGURE 3-43 SEGMENT OF FAIRFAX, VA
1:50,000 SCALE
USGS

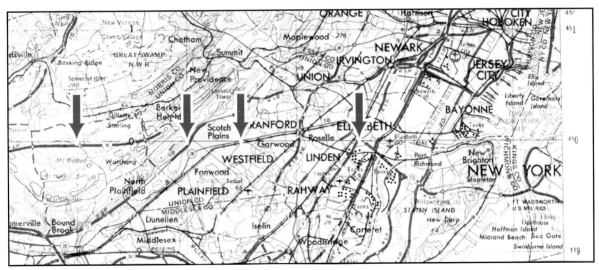

FIGURE 3-44 SEGMENT OF NEWARK, NJ, PA
1:250,000 SCALE
USGS

map using the "common points" technique described above (**Figure 3-44**). Use care in drawing these lines to insure as much accuracy as possible. But remember, we are not going to use them as part of a legal survey—we plan to use them to help us find our way. Minor imprecisions should not cause us to become lost.

FIGURE 3-45

Municipal and county street and road maps can be prepared in this same fashion by local governmental and volunteer public safety and emergency response organizations. Perhaps your county map cartography house will begin to include this MGRS/UTM grid coordinate information on future editions of your maps, if you explain that you require it. Regardless, the time spent in preparing your maps with MGRS/UTM grid information will pay large dividends in terms of the quality and speed of public safety services provided to your community.

The necessary base maps for accomplishing this map preparation work can be obtained in many local book and map stores or by contacting the appropriate governmental agency. For example, large-, medium-, and small-scale topographic maps, as well as free map indexes and catalogs for each state in the U.S.A. (**Figure 3-45**) and an index to those numerous military map sheets produced by the DMA available to the general public, are available from the National Cartographic Information Center (NCIC),U.S. Geological Survey, 507 National Center, Reston, VA, 22092 (1-800-USA-MAPS). Direct contact can also be made with the Defense Mapping Agency Combat Support Center, Attn: PMSR, Washington, DC, 20315 (202-227-2495). Many states have map production and information units, as well. Finally, several libraries throughout the USA have been designated as **Map Depository Libraries** (e.g., The Lafayette College Geology Library in Easton, PA,USA) for many of the published maps of the U.S.G.S. and each is listed in the **Catalog of Published Maps** for each state.

For maps of areas within Canada, contact the Canada Map Office, 615 Booth Street, Ottawa, Ontario K1A 0E9 (613-952-7000). One of the best single sources for information on obtaining maps world-wide is the third edition of <u>**The Map Catalog: Every Kind of Map and Chart On Earth and Even Some Above It**</u>, by Joel Makower, editor. It was published in 1992 by Vintage Books, a division of Random House, Inc., New York.

Finally, you can position new features and correct mistakes on existing maps by accurately determining a location using your GPS receiver. For example, a new trail or highway can be mapped by obtaining a series of position fixes along its course, plotting these points on the map, and then connecting the dots with the line being added to represent it. Point features, such as the newly constructed Farinon Center at Lafayette College, can simply be added to the Easton, PA,NJ, USA, map using the coordinates determined by standing there with the GPS unit (18TVA825 052).

After this preparation work is completed, the map you have selected is ready for use with your GPS equipment.

THE LANGUAGE OF MAPS

No one knows who first drew, molded, laced together or scratched out in the dirt the first map. A map useful for navigation is a graphic representation of a portion of the earth's surface drawn to scale, generally as seen from directly above. It portrays two types of features: (1) **natural features**, which may include water (hydrography), vegetation, and relief (topography), and (2) those **cultural features** erected by man. Depending upon the map's scale and intended purpose, these cultural features included may be streets and highways, bridges, railroads, power and communications transmission lines, and buildings. There may also be included the various political boundaries that exist in the area portrayed.

Now it is time to begin the task of reading the map. Maps convey information to you in four ways: (1) marginal information, (2) colors, (3) symbols, and (4) labels.

MARGINAL INFORMATION

Items of marginal information are generally classified into three categories: (1) map identifications, (2) map interpretation and use, and (3) other miscellaneous data. **Map identifications data**, such as sheet names and edition numbers, series names and numbers, MGRS grid reference boxes, scale notes, credit notes, adjoining sheets diagrams, and stock numbers all help you to determine the location and amount of area covered by a map and to identify the specific sheet you are examining. **Interpretation and use data**, such as compilation dates, legends showing the meanings of the colors and symbols used on the map, magnetic declination diagrams, contour intervals, and graphic distance scales, all assist the reader to fully understand and utilize the map. Finally, **miscellaneous data** might include such ancillary information as an index to boundaries, mileage chart between cities, and so forth.

It should be noted that not every map includes all the marginal information discussed here. Civilian and military topographic maps, however, will include much of this information.

COLORS AND SYMBOLS

In the development of the language of maps, there has been an attempt over the centuries to apply logic to the process. This is true for the use of both colors and symbols.

By the fifteenth century, most European maps were carefully colored. Profile drawings of mountains and hills were shown in brown, rivers and lakes in blue, vegetation in green, roads in black or yellow, and special information in red. As we can see, the use of colors hasn't changed much since that time. A quick check of the map's **legend** will help you to confirm whether or not the map you are using conforms rather closely to the age-old color scheme described above. Most likely it does.

One difference may be that major highways are shown in reddish brown on U.S. topographic maps and are generally shown in red and black on U.S. commercial highway maps. On the other hand, Michelin highway maps of Europe, for example, show them in red, yellow, and black—in order of their importance.

The legend, again, is the place to look when searching for the meanings of various colors and symbols found on any particular map (Figure 3-46). This is true whether you are using a civilian or military topographic map produced by a government agency or a commercially produced street or highway map. Most symbols are obvious, but some may be somewhat

FIGURE 3-46

USGS

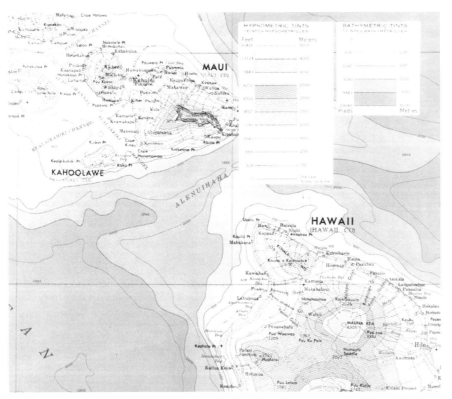

FIGURE 3-47 SEGMENT OF HAWAII 1:1,000,000 SCALE INTERNATIONAL MAP OF THE WORLD

difficult to associate with the features they represent. Due to maps' scale restrictions and legibility considerations, symbols placed on them are often greatly exaggerated in size. For example, a road 20 meters wide may be shown on a 1:50,000-scale topographic map as a line that would, according to the map's scale, measure 60 meters in width. As a result, some features must also be displaced in order to depict them (e.g., a highway running parallel to a stream and railroad track) and some may be left out (e.g., some buildings within a cluster or along streets within a heavily built-up urban place) because these symbols are exaggerated in size. This is why the user must interpret the map—not just read it.

If you have a physical map, it may use elevation tints (various colors or shades of black and white) to identify the various elevation intervals set for that sheet (Figure 3-47). Or, it may use line symbols, called contour lines (lines connecting points of equal elevation), to convey specific elevation information as well as portray local relief (Figure 3-48). Relief is the shape (topography)

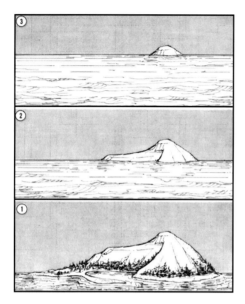

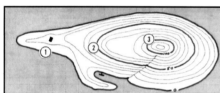

FIGURE 3-48

of the local surface area being portrayed (Figure 3-49). Contour lines were first used on a map to portray elevation changes in 1749.

The vertical interval (elevation difference) between contour lines on the map is referred to as the map's **contour interval**, which is reported as a marginal note. Generally, areas with pronounced relief differences (i.e., mountains) have larger contour intervals than relatively flat areas. However, the greater the contour interval used on a map, the less specific will be the portrayal of the physical shape of the land in its relatively flat areas. Thus, more interpretation will be required by you, the reader. There will be some hints about how to do this later in the chapter.

Today's topographic maps generally use only four types of contour lines to depict the infinite number of configurations the land can take. They are:

Index Contour Lines - Every fifth contour line, which is drawn a bit heavier than the four lighter ones in-between, is called an index contour line. These lines are periodically broken so their elevations can be printed on the map (**Figure 3-50**). Incidentally, the top of this elevation number (**Figure 3-51**) is **usually** (right circle) – but not always (left circle) – pointing upgrade .

Intermediate Contour Lines - The four lighter contour lines located between any two index contours are called intermediate contour lines (**Figure 3-50**). Remember, the vertical distance between individual contour lines (either between an index and intermediate

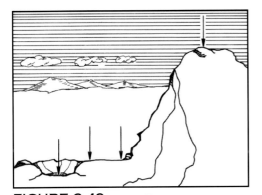

FIGURE 3-49　LOCAL RELIEF

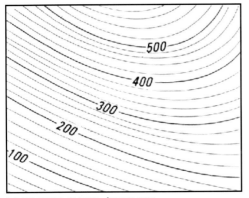

FIGURE 3-50　INDEX AND
INTERMEDIATE CONTOUR LINES

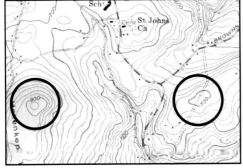

FIGURE 3-51　NUMBER TOPS USUALLY
POINT UPGRADE

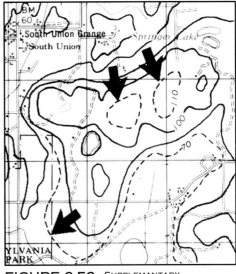

FIGURE 3-52　SUPPLEMANTARY
CONTOUR LINES

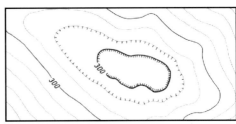

FIGURE 3-53　DEPRESSION
CONTOUR LINES

or between intermediate contour lines) is always equal to the contour interval. Elevation values are generally not printed on intermediate contour lines.

Supplementary Contour Lines - On some maps, particularly in those specific areas of the map where there is little local relief, there may appear either very light or broken contour lines. They are placed there to

FIGURE 3-54

STEEP SLOPE

FIGURE 3-55 GENTLE SLOPE

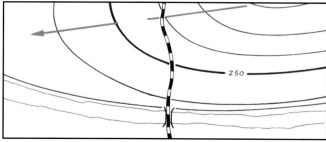

portray the shape of the land which might not otherwise be perceivable because its contour interval is too large to effectively portray the topography of these areas (**Figure 3-52**). They generally, but not always, represent half the map's contour interval and their elevations are frequently labeled.

Depression Contour Lines - An area such as a gravel pit, a man-made cut to accommodate a highway, or some form of natural depression that is of lower elevation than all the immediate surrounding terrain is portrayed by use of depression contour lines. They have small hachures (short dash lines) pointing down slope, and each may be labeled with its elevation (**Figure 3-53**).

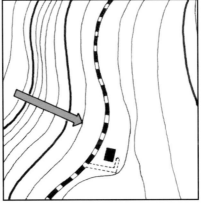

Every wiggle (twist or turn) in your map's contour lines, no matter what size, signals what is to be encountered while traveling across the space that falls between them. Relatively wide spaces between contour lines are rarely flat places or areas of uniform slope, but the only way to tell for sure is to look for minute bends or wiggles in the contour lines found on either side. In fact, it is wise to look at the next two or three lines to help you determine just what you will encounter there on the ground.

FIGURE 3-56

CONCAVE SLOPE

As you will recall, the closer the contour lines are placed together (the less horizontal distance falling between them), the greater is the slope being portrayed (**Figures 3-54 & 3-55**). Also, when a slope is steeper at the top (contour lines are closer together) than at the bottom (where the lines are further apart), the slope is **concave** in shape (**Figure 3-56**). Conversely, when the slope is more gentle on the top (contour lines are further apart) and steeper at the bottom (contour lines are closer together), the slope is convex in shape (**Fig-**

FIGURE 3-57

CONVEX SLOPE

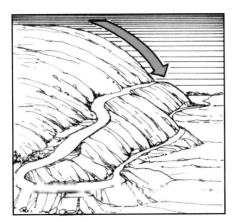

FIGURE 3-58 Uniform Slope

FIGURE 3-59

ure 3-57). Finally, when the contour lines are evenly spaced, the slope is **uniform** in shape (**Figure 3-58**). To better understand these fundamental concepts for interpreting the shapes of slopes using contour lines on a map, see **Figure 3-59**.

Geologists and geographers have developed thousands of terms to define and describe the various landforms found out in the real world, but we are simply trying to observe, interpret, and recognize what nature displays as a guide to our movements. Therefore, our classification list of five **major terrain features**, two **minor terrain features**, three **supplementary terrain features**, and the concept referred to as a **complex terrain feature**, is quite sufficient for our purposes.

The **five major terrain features** are (1) hill, (2) ridge, (3) saddle, (4) valley, and (5) depression. A **hill** tends to be round (although not perfectly) and slopes downward in all directions. A **ridge** is an elongated piece of high ground, generally with three downward and one upward slope along its crest. **Saddles** are either wide or narrow and deep or shallow dips between hilltops or along the crest of a ridge. A **valley** slopes rather steeply upward in two directions and gently upward and downward in the other two directions. Finally, a **depression** is either a natural or artificial hole in the ground (e.g., a sand or gravel pit) with the ground sloping downward in all directions toward its generally wet center.

The two **minor terrain features** are (1) spur and (2) draw. Basically, a spur is a small ridge and a draw a small valley. Often, they are proportionately steeper than their larger counterparts.

The three **supplementary terrain features** are (1) cliff, (2) cut, and (3) fill. A cliff can be nearly vertical. Its portrayal on the map features contour lines placed either very close together or merging into a single line denoting a sheer, perpendicular slope. Cuts and fills are created by man to prepare the ground for the bed of a highway or railroad in order to reduce the slope or keep the roadway dry through low, swampy areas.

FIGURE 3-60

A. CLIFF   B. HILL   C. SPUR   D. DRAW
E. VALLEY   F. RIDGE   G. SADDLE
H. DEPRESSION   I. FILL   J. CUT

Figure **3-60** illustrates how each of these terrain features might appear in the real world. **Figure 3-61** helps you to see how contour lines are used on a map to represent various slopes, shapes, and features found on the earth's surface. And, finally, **Figure 3-62** presents a USGS illustration of how the terrain of a hypothetical area would appear as contour lines on a large-scale topographic map.

You may have noticed that you will generally not find terrain features standing alone. They are most often part of a larger, more **complex composite terrain feature** that stands out as a single observable entity.

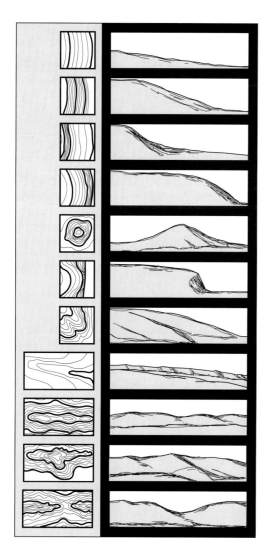

FIGURE 3-61

FIGURE 3-62   USGS

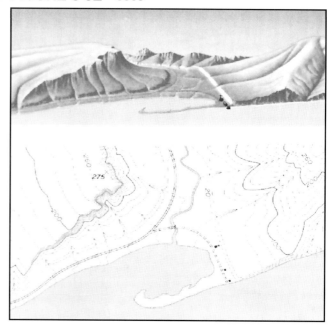

Ken White, who served for several years as the Land Navigation Instruction Chief at the Committee Group, U.S. Army Armor School, Fort Knox, KY, USA, developed a strategy that is quite useful for interpreting contour lines on a topographic map. When interpreting microrelief—those features not shown directly by contour lines due to the size of a map's contour interval—you can make use of the implications made by the presence of their small wiggles and larger curves. He believes that map interpretation is a matter of learning to "read between the lines." **Figures 3-63 a.-d.** explain much with few words. **Figures 3-64 & 3-65** also illustrate how to read between the contour lines by closely observing either their pronounced curves or more subtle "wiggle" patterns found on a map.

DESCRIPTIVE LABELS

Colors and symbols are the two most fundamental tools used by cartographers to convey information on their maps. However, reading **descriptive labels** can add much to your understanding of what is portrayed. They are used to identify the names of cities, towns, roads, mountains, rivers, and so forth; but they are also used to more specifically describe general symbols found on the map. For example they may identify oil tanks, post office buildings, apple orchards, fire towers, rice paddies, and so forth.

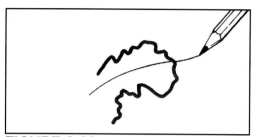

FIGURE 3-63A　Crest of Small Spur

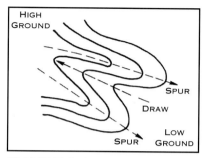

FIGURE 3-64

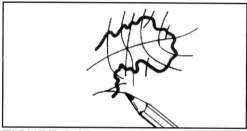

FIGURE 3-63B　Minor Crests Added

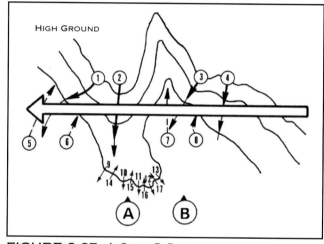

FIGURE 3-65　A. Spur B. Draw
1 - 4, Small Spurs; 5 - 8, Small Draws;
9 - 13, Micro Spurs; 14 - 17, Micro Draws

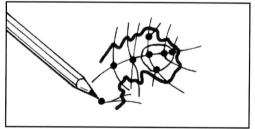

FIGURE 3-63C　Small Hills
Where Crests Converge

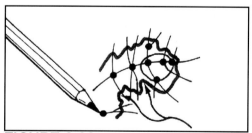

FIGURE 3-63D　Small Draws
Between Minor Crests

SELECTING A GOOD ROUTE

Much of your success or failure may be determined before you ever take that first step or roll over that first mile or kilometer toward your destination. As for any type of activity, the foundation for success in navigation rests upon the quality of thought and effort you devote to planning and preparing for your move.

When selecting a route, it should be done in conjunction with a study of the map. Simply asking your GPS unit to give you the direction and distance from your starting point through several intermediate checkpoints and on to the final objective can lead to many serious problems. This approach completely ignores the realities presented by the terrain, weather-related ground conditions, and the numerous special considerations placed upon the move by the task at hand. In addition, you must consider the number of people and types of equipment to be moved, and the time available to get there and complete the task.

Don't forget Forest Ranger Hagedorn's caution that we consider the terrain from one point to another in terms of the time, effort, and difficulties to be encountered when moving over it (functional distance), rather than only in terms of linear distance. Thus, you must carefully study your map and expect this study to result often in the planning of complex multi-segmented routes. It should now be obvious that the GPS unit's capability for storing and helping you to navigate over a preprogrammed route with multiple legs will be quite beneficial.

A map study focused on the trail, street, and highway networks of an area to be traversed is just as important to the success of a road movement as it is to cross-country movement. Many of the same considerations must be reviewed prior to making routing decisions in either case. For example, heavy rains often wash out dirt roads and trails in hilly and mountainous areas, bridges on secondary roads often do not support heavy loads, and travel times on level roads over greater distances may be less than those over steep, twisting "shortcuts."

KEEPING TO THE ROUTE

After selecting a good route to your destination, effective use of your map and GPS can keep you on it with little difficulty. Periodic position checks will tell you if or when you stray from the route and the map will encourage you to use the terrain and other natural and man-made features to help direct and guide your movement between these position checks.

Position information from your GPS receiver is not particularly useful until it is placed within the context of the reality confronting you on the ground. We are not really interested in going to a set of coordinates; we wish to proceed to a real world place represented by those coordinates (e.g., base camp, a fishing hole, the place we plan to meet for lunch or Wayne's house for a party).

As we have already stated, straight line travel over moderate and long distances is impractical because the terrain, highways, and numerous other factors preclude it. For example, the terrain will channel and guide both your route selections and movements. Therefore, being able to positively identify the terrain and other features you encounter is of great importance to your navigational success.

In fact, the ability to identify specific terrain features simultaneously on the ground and on the map can be extremely helpful as you proceed over the land. Then, by using an association between the terrain and the map, you can mentally "hand-off" your position from feature to feature as they guide you almost effortlessly along the route toward your destination. Whenever the level of uncertainty about your position mounts, you can confirm it with GPS and make any necessary correction. In other words, use of the a GPS receiver in conjunction with map-terrain referencing skills cause this movement technique to become what will be known as **"fail-safe terrain association"**.

Here is a simple strategy you might employ for recognizing and identifying specific terrain features encountered on the ground and portrayed on the map. It was first developed by the U.S. Navy Personnel Research and Development Center (NPRDC) for the U.S. Marine Corps and was later refined by the U.S. Army Research Institute for the Behavioral and Social Sciences (ARI). A modified version **Map Interpretation Terrain Association Course (MITAC)**, by Jennifer N. Drescher, was developed by Alexis USA, Inc. and is presently in use for training military personnel on the techniques of movement by terrain association in the Kingdom of Saudi Arabia. Soldiers attending the "Light Fighter School," 10th Mountain Division (Light Infantry) at Fort Drum, NY, USA, are also using this technique.

First, this terrain assocication technique requires that you be able to name and identify the ten classifications of terrain features presented earlier. They are hill, ridge, saddle, valley, depression, spur, draw, cliff, cut, and fill. Once a particular feature has been classified, it can then be identified both on the map and in the real world by analyzing it in terms of up to five of its physical characteristics known by the acronym SOSES. These physical characteristics include: (1) shape, (2) orientation, (3) size, (4) elevation, and (5) slope.

Identifying Specific Terrain Features Using Five Physical Characteristics (SOSES)

- **Shape** - the general form or outline of the feature at its base. It may be (1) **round** or (2) **elongated**.

- **Orientation** - the general trend or direction of an elongated feature from your viewpoint. A feature can be (1) **in line**, (2) **across**, or (3) **at an angle** to your viewpoint.

- **Size** - the length or width of a feature horizontally across its base. For example, one landform might be (1) **larger** or (2) **smaller** than another.

- **Elevation** - the height of a landform. This can be described either in absolute terms or as compared to other features in the area or to your own position. One landform may be (1) **higher**, (2) **lower**, (3) **deeper**, or (4) **shallower** than another.

- **Slope** - the type and steepness of the slope on either side of the landform (left or right). These slopes may be (1) **uniform**, (2) **convex**, or **(3) concave** and they may be (1) **steep** or (2) **gentle**.

FIGURE 3-66

FEATURE AND 1:50,000 SCALE MAP SEGMENT NEAR AL JUMUM, KSA

It may not be necessary or even possible to use all five physical characteristics in identifying all landforms. You are to use only as many of them as may be useful or necessary in allowing you to make positive identifications. You will find this technique invaluable in generally keeping track of your position on the map between more precise position fixes with the GPS receiver and as a means for guiding you along the route over and around the various features being encountered (**Figure 3-66**).

Other terrain referencing techniques that you might find helpful in guiding your movements are the use of (1) **handrails**, (2) **catching features**, and (3) **navigational attack points**. Navigational **handrails** are linear features, like roads or highways, railroads, power transmission lines, ridgelines, or streams, that run roughly parallel to your direction of travel. Instead of paying close attention to compass readings and position fixes, you can proceed quickly using the "handrail" as a guide to your movement.

When you reach the point where your route and the handrail are to part company, you can make use of a prominent feature located nearby to serve as a warning. This feature is now being used as a **catching feature**. You may also use your GPS unit's navigation capability along each route segment to notify you when you have arrived at the next checkpoint, which you have pre-planned to also serve as a catching feature. A similar feature may also serve as your **navigational attack point**, but that discussion fits better under the next heading.

RECOGNIZING THE OBJECTIVE

A **navigational attack point** is a readily-identified feature located near your final objective when the objective is rather obscure or difficult to recognize. From this attack point, a short precision movement, carefully negotiated with a compass and monitored with your GPS receiver, will then easily bring you to your final objective or destination. By using the map and the terrain, referring to your compass (next chapter), and employing the navigational capabilities of your GPS receiver (Chapter 5), you will have no difficulty either in keeping to the route or recognizing the objective.

IN SUMMARY

You should now be able to read and interpret maps to the extent necessary for you to use them effectively as you navigate with GPS. With a map in hand, you can know any area as if it were part of your own neighborhood. You can grasp the complex interrelationships among the several places, various routes, numerous features, and the positional grid found printed on it. Thus, with GPS, you will always know where you are and where you are going in relation to: (1) the map's grid, (2) the map's feature portrayal, and (3) the complete inventory of real world features to be encountered on the ground. In summary, with the marriage of the map and GPS, your mental navigational and locational perceptions will now very closely match reality.

When using maps, such as topographical maps showing a great number of natural and cultural features, you should recall that it is the terrain and hydrographic (water) features that serve as the most reliable movement guides because they change least over time. Both vegetation and those features constructed by man are frequently changing. In addition, map scale, compilation date, and contour interval have a great impact upon what is and is not shown on a map. Maps are not photographs or comprehensive portraits of what is found on the ground—they are only partially complete graphic tools for the navigator. Remember, too, the numerous pieces of information contained in the map's margin, especially the legend, and the labels included along with the map's symbolic portrayal can add greatly to your level of understanding, if you take the time to study them in detail.

Where few features of any type exist, most will be included on the map (e.g., buildings, water, vegetation, and trails in the desert). And, where many exist, several will be omitted (e.g., stream courses in well-watered areas and buildings and streets in urban settings). You must also recall that many feature symbols are exaggerated in size on the map to insure legibility. Also, vegetation on the ground may mask more than it reveals.

Finally, you must orient your map to the ground (within 30°) to effectively use it. Techniques for orienting the map will be explained in the next chapter.

Khabt ad Di'aythah

GPS NAVIGATION WITH AND WITHOUT A COMPASS

(SPATIAL RELATIONSHIPS: DIRECTION & DISTANCE)

This chapter will explain how to integrate to maximum advantage the use of a magnetic compass along with the GPS receiver and a map as you proceed to find your way. Even during this age of GPS technology, your compass remains the most useful tool for determining direction out on the ground and for orienting your map.

Whenever you think about determining direction, you must also consider the measurement of distance because of their common spatial link. All features and positions on the earth's surface are held in a unique spatial relationship—each has a specific direction and distance to and from all the others. In other words, when moving from here to there, you must know which way to go and how far. Thus, a discussion of distance will also be included in this chapter.

BACKGROUND

Although you can use the navigational feature on a GPS unit to determine what direction you are moving and keep correcting yourself through trial and error, it is the magnetic compass that will most quickly and easily point you in the right direction out on the ground. You can then use the GPS unit to help keep track of your progress and suggest steering corrections as you proceed along your route.

The military lensatic compass will report directional azimuth readings in either degrees (360 parts of a circle) or mils (6400 parts of a circle), but it is suggested that you use degrees for LN purposes.

Although nearly every type of magnetic compass will work, your best choice is use of the lensatic compass meeting current U.S. military specifications (**Figure 4-1**). It is designed to withstand rugged field conditions, uses copper induction damping, is most accurate in sighting directions for targeting and navigational purposes, and contains vials of tritium gas that continuously illuminate its nighttime features. The military style lensatic compass is manufactured by Stocker & Yale, Inc. in Salem, New Hampshire, U.S.A.

In comparison, popular liquid damped protractor-base style compasses have an upper temperature range of only 120° F before their capsules begin to fail (crack and leak) due to expansion. Also, they are not as accurate when precise sighting is required. And, finally, they are difficult to use in the dark. For more information regarding the advantages of using the military lensatic compass, see "In Defense of the Lensatic Compass," by N.J. Hotchkiss, on pages 31-34 of the November-December 1991 issue of **Infantry**, a bimonthly professional military publication of the U.S. Army Infantry School at Fort Benning, GA, USA, 31905-5593.

HOW MANY NORTH S ARE THERE?

Before we embark on a discussion about determining direction on both a map and on the ground, you should understand that map and compass norths are generally not the same direction. This condition is known as **magnetic declination** (sometimes called **compass variation**).

There are, in fact, three norths on any map (**Figure 4-2**): true north (★), grid north (**GN**), and magnetic north (*I*). **True north** is the direction you would take to travel to the North Pole and **grid north** is the direction represented by going straight up a grid line toward the top of a map. Land navigators need not concern themselves with the small difference between true and grid norths, which results when map makers represent a part of the spherically shaped earth on a flat piece of paper. How-

MAGNETIC POLE   GEOGRAPHIC NORTH POLE

FORT DRUM

GEOGRAPHIC SOUTH POLE   MAGNETIC POLE

GN ★

12°

1/2°

DECLINATION DIAGRAM
FORT DRUM, NY

ever, they must consider the difference between **grid north** (map north) and **magnetic north**, the direction in which the compass needle points. The variation between them is called the **grid-magnetic (G-M) angle.**

TRUE NORTH

THE DIRECTION FROM ANY POSITION ON THE EARTH'S SURFACE TO THE NORTH POLE. ALL LINES OF LONGITUDE ARE TRUE NORTH LINES. THIS REFERENCE POINT IS SYMBOLIZED BY A STAR.

GN

GRID NORTH

THE NORTH THAT IS ESTABLISHED BY THE VERTICAL GRID LINES ON THE MAP. THIS REFERENCE POINT IS SYMBOLIZED BY THE LETTERS GN.

MAGNETIC NORTH

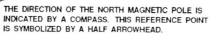

THE DIRECTION OF THE NORTH MAGNETIC POLE IS INDICATED BY A COMPASS. THIS REFERENCE POINT IS SYMBOLIZED BY A HALF ARROWHEAD.

FIGURE 4-2

Magnetic declination results from the fact that the compass needle is attracted by the earth's magnetic force patterns that presently converge in the northern hemisphere at a location near Bathurst Island in northern Canada. In the contiguous United States, the G-M angle presently varies from 20° west of grid north in parts of Maine to 21° east in parts of the State of Washington. There are, of course, places in the world where there is little or no declination. For example, there is a line, called the Agonic line, in the U.S.A. along which there is no magnetic declination. It lies just off the west coast of Florida; runs up through Georgia, Tennessee, and Kentucky; on along the western shore of Lake Michigan; and into Canada. The Middle East, too, is an area with little magnetic declination.

Although GPS receivers can be set to report already converted magnetic azimuths among stored positions (waypoints), you must consider this mathematical conversion requirement between grid and magnetic azimuths when you measure directional values with a protractor directly on a map.

ORIENTING THE MAP

Don't forget the paramount rule for LN: **Always navigate with a correctly oriented map.** The amount of error should never exceed 30°. Use of a lensatic compass is the quickest and easiest method for accomplishing this task, but inspection (terrain referencing) can also work in the absence of a compass.

ORIENTING THE MAP USING A LENSATIC COMPASS

Just follow this simple two-step method:

1. With the map laying flat, place the compass on it parallel to a north-south grid line with the cover end of the compass pointing toward the top of the map.

2. Rotate the map (and compass) until the declination angle formed by the black index line and the compass needle match the declination diagram (or note) printed in the margin of the map. The map is now oriented (**Figure 4-3**).

Be certain the magnetic arrow is resting on the correct side (east or west) of the index line, as compared to the declination diagram that may be found in the

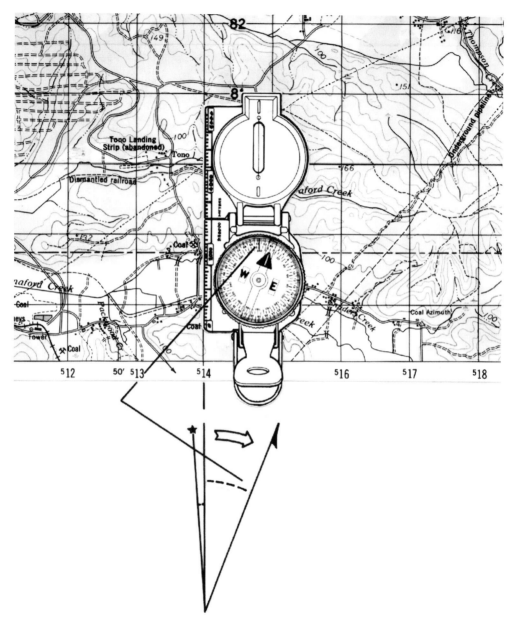

FIGURE 4-3

margin of your large-scale topographic map. You should be able to accomplish this task to an accuracy of within 3°, but the map should never be misoriented to the ground more than 30° in order to avoid almost certain confusion.

When the map you are using has no diagram or marginal note describing the area's magnetic declination, you can obtain that information for any geographic area by consulting the appropriate large- or medium-scale topographic map (e.g., USGS or DMA quadrangle sheets). When this is not practical, better than 30°

FIGURE 4-4

accuracy can normally be achieved by laying the compass along the edge of any map and orienting it to magnetic north. If more accuracy is desired, you can stand on or near a linear feature, such as a road, facing in the direction which is known from its portrayal on the map. After taking a compass reading, the magnetic declination for the area in which you are located will become apparent as you compare the reading to the measured directional azimuth taken by using a protractor on the map.

ORIENTING THE MAP USING INSPECTION (TERRAIN REFERENCING)

Follow these three steps to orient the map to the ground by map-terrain referencing:

1. Find your position using the GPS.

2. Look at three or four features on the ground in common with the portrayal on the map, such as hilltops, saddles, valleys, ridges, highways, buildings, bridges, or rivers and streams.

3. Rotate the map as you inspect it for these features until its portrayal is aligned with these same features in the real world (**Figure 4-4**).

DETERMINING DIRECTION AMONG POINTS ON A MAP

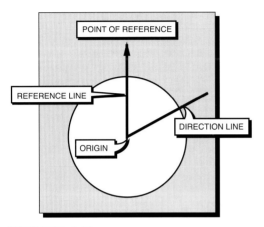

FIGURE 4-5

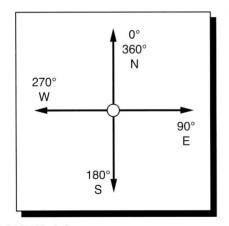

FIGURE 4-6

The **best** method for determining the direction (and distance) between any two positions is to save as waypoints their coordinates and elevations from a map into the memory of the receiver and then let the unit calculate these values from one to the other. GPS units can be set to report these azimuth directions either in magnetic or true values. For navigational purposes, it is best to have the unit set to give you magnetic azimuth values so they can be used in conjunction with your lensatic compass without first making conversion calculations.

An azimuth is any directional value being measured as an angle that is read in a clockwise direction from a north (0°) reference line (**Figure 4-5**). For example, north is at a 0° or 360° azimuth, east at a 90° azimuth, south at 180°, and west at 270° (**Figure 4-6**).

There are times when you may wish to measure directional values directly on a map. For example, you may be required to make some short, finely tuned directional changes along a route between major checkpoints when an obstacle or some other unexpected difficulty is encountered on the ground in order to make the movement safer or easier.

Determining direction directly from a map is best done through use of a 360° protractor with 0° on the

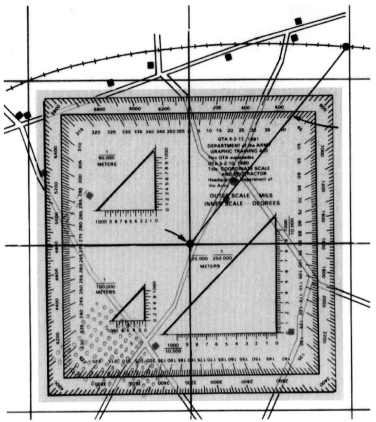

FIGURE 4-7

scale being oriented toward grid (map) north. When using this method, the next illustration, **Figure 4-7**, shows the correct method for placing a square military-style protractor on the map and measuring directional values. Note that the center of the protractor is placed over the position from which the azimuth is to be measured and the index lines on the protractor are oriented parallel to the grid lines on the map.

When measuring a directional azimuth directly on the map, don't forget to convert the grid azimuth values being read on your map before applying them to your compass. Many large-scale topographic maps include conversion notes in their margins that are quite useful when you make these conversion calculations. **Figure 4-8** contains the conversion notes from the (a) TENINO (Fort Lewis), Washington, USA, 1:50,000-scale map sheet and (b) the FORT DRUM, New York, USA, 1:50,000-scale map sheet.

Incidentally, there is a protractor-like tool available from Stocker & Yale, Inc. called a Declitractor®, that properly offsets the 360° protractor scale by first setting the G-M angle on the device. This instrument can plot and report all azimuths on the map in magnetic values without conversion. (**Figure 4-9**).

DETERMINING DIRECTION ON THE GROUND

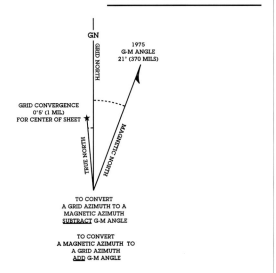

FIGURE 4-8A

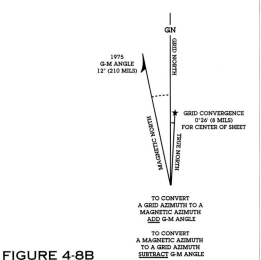

FIGURE 4-8B

The military lensatic compass (**Figure 4-10**) is the best instrument for determining direction on the ground. It contains a "floating" protractor scale with a mounted (magnetic) north-seeking arrow that measures the angular values for any direction (azimuth) in the real world.

You will note that the dial of the compass, known as the compass card, has an inner (red) protractor scale printed on it with the north-pointing arrow being defined as 0° or 360°. Just as on the map protractor, north on the compass is at a 0° or 360° azimuth, east at 90°, south at 180°, and west at 270° (**Figure 4-11**). Note that a black mil scale is located on the outer edge of the compass card. Remember, it is the red (inner) degrees scale, marked off in 5° increments, that is used in LN.

When using a lensatic compass, you read directional azimuth values from the protractor scale directly under the black index line printed on the cover glass. There are two methods for holding the compass for daylight use: (1) the center-hold technique (**Figures 4-12 & 4-13**) and (2) the compass-to-cheek method (**Figures 4-14 & 4-15**), which is the most accurate.

For night use, the luminous line and bezel ring serration and clicking device are invaluable. Starting with the luminous line oriented right over the black index line, for every counterclockwise click of the bezel

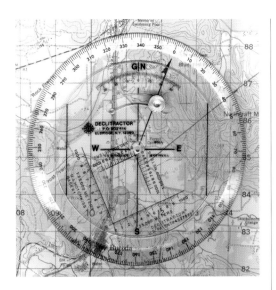

FIGURE 4-9   DECLITRACTOR
A DEVICE QUICKLY BECOMING A NECESSITY FOR
ANY SERIOUS WORK WITH MAPS

FIGURE 4-10

ring, the luminous line is moved +3°. To set the compass for use at a particular azimuth at night (for example 30°), the luminous line is set over the black index line and then rotated 10 clicks (30° / 3 = 10 clicks) counterclockwise. Now, holding your open compass so that it is pointing directly out in front of you, rotate yourself until the luminous north arrow is lined up with the luminous line. You are facing at an azimuth direction of 30° (**Figure 4-16**).

There is a quick method for presetting the compass during daylight for night use. Place your open compass in the center-hold position and face in the desired direction for the preset (i.e., 275°). While holding the compass steady with a reading of 275° under your black index line, rotate the bezel ring so that the luminous line aligns with the luminous north-seeking arrow. Later, when you open the compass and rotate yourself to again align the north arrow with the luminous line, you are facing in the desired azimuth direction of 275°.

Before using a lensatic compass, be certain all the parts are there and functioning properly. Also, do not use your compass near high tension power lines (55 meters), truck (20 meters), telephone or barbed wires (10 meters), or a rifle (.5 meters). Metal eyeglass frames may also affect the accuracy of a compass reading when using the compass-to-cheek method.

Yes, the lensatic compass can and should be used during vehicular mounted navigation as well.

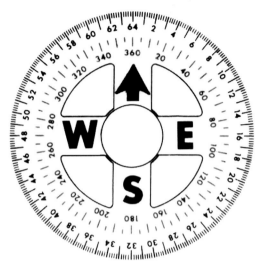

FIGURE 4-11 LENSATIC COMPASS CARD

There is a five-step method for using it on a tactical or any other type of vehicle (**Figure 4-17**):

1. Dismount and move at least 25 meters forward of the vehicle.

2. Locate a distant sighting point and take a directional reading with your compass.

3. Remount the vehicle and move slowly forward toward the distant sighting point and take another reading.

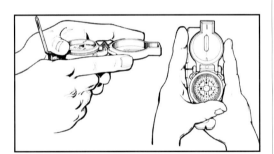

FIGURE 4-12

FIGURE 4-13 CENTER HOLD TECHNIQUE

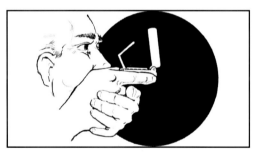

FIGURE 4-14

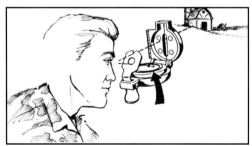

FIGURE 4-15 COMPASS-TO-CHEEK TECHNIQUE

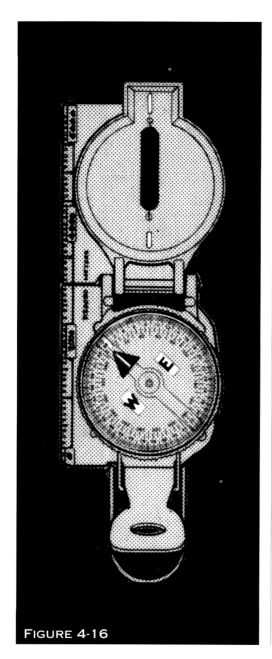

FIGURE 4-16

4. Determine the difference between the two readings, which is the error (compass deviation) caused by the metal and electrical system within the vehicle. This deviation is generally less than + or - 10°.

FIGURE 4-17

5. Apply that deviation (add or subtract) to every subsequent reading you take from this vehicle using that compass.

In summary, whenever you navigate over the land, you should have a magnetic compass to use in conjunction with your GPS receiver and map. The GPS receiver locates your position by coordinates, the map relates this position to your surroundings through its grid and feature portrayal, and the compass orients both you and the map to these surroundings by reporting real world directions. These are the three vital ingredients needed for that first important step in LN – **know where you are** in every possible way: (1) on the map, (2) on the ground, and (3) in relation to all features and positions in your area.

FIELD EXPEDIENT METHODS FOR DETERMINING DIRECTION ON THE GROUND

When you don't have a compass, there are some techniques you can employ for determining direction. Here are just a few.

USE OF GPS

You can use your receiver's graphic steering feature to determine direction at any time. Establish a route from your present position to some other point on the map after having already saved both as waypoints. Set out in any direction and, using the navigation feature, the unit will determine the direction you are actually moving. With that accurate directional information, you can then estimate any real world direction from your position.

Just be careful you don't face in a different direction during a halt because the unit shows your direction of travel in relation to your forward progress. **It is not a compass**. If you turn your body while standing in place, you will misorient the graphic display

DAYTIME EXPEDIENT TECHNIQUES

1. At noon (standard time), the sun is nearly due south for most of the northern hemisphere and due north for most of the southern hemisphere. Remember that this will not apply in tropical areas.

2. The sun melts snow more quickly on south-facing slopes in the northern hemisphere and on north-facing slopes in the southern hemisphere. This may be obvious during winter months, but it may also be determined during the summer because these sun-facing slopes experience more erosion due to water runoff during the winter when vegetation is less able to hold the soil in place.

3. Sunflowers face east.

4. Blue sky over large bodies of water may appear darker than sky over land. At times, you may be able to guide your movement by this darker sky along a coastline or large lakeshore.

5. An oasis may develop a cloud that can serve as a locational and directional guide.

6. Certain areas have prevailing winds, such as the "westerlies" over much of the continental United States. They may be helpful in roughly judging direction, especially over short periods of time if the wind direction was checked before starting the movement.

7. A quick look at a map reveals prevailing drainage patterns or directions in which ridgelines or hill or mountain chains are generally oriented in an area. These natural cues can be useful if attention is paid to them during the map reconnaissance prior to the start of a movement.

NIGHTTIME EXPEDIENT TECHNIQUES

1.   You can tell direction from the stars.

a.   In the **northern hemisphere**, locate the Big Dipper constellation in the sky (**Figure 4-18**). The two front stars of the "soup ladle" point almost directly toward the North Star (Polaris), which is about five times the distance between these two front pointer stars above the open top of the dipper. When you face the North Star, you are always facing to within 58 of true north.

b.   **In the southern hemisphere**, locate the Southern Cross star pattern in the sky (**Figure 4-19**). Although there is no star located in the sky directly above the South Pole, this imaginary south point in the sky can be easily determined. It is 4 1/2 times the distance from the top to the bottom of the Southern Cross and below it. You should then select a landmark directly below this imaginary "south spot" in the sky and face toward it to estimate any other direction on the ground.

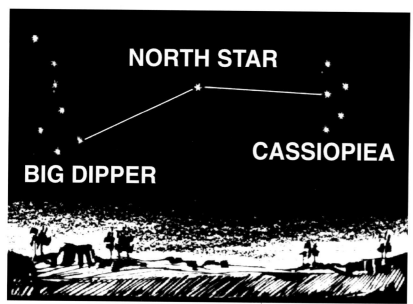

FIGURE 4-18

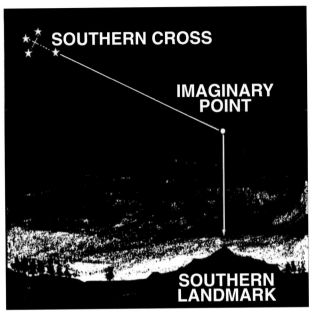

FIGURE 4-19

2. You can sometimes tell direction by observing the moon. An imaginary line drawn through the "horn-tips" of a crescent moon will run approximately north and south.

3. A nighttime rural sky is dark, but an urban sky is bright and will serve as a navigational beacon that can be seen for many miles. Notice the bright "blink" in the sky above even small urban places.

4. Smells and sounds can also help you keep track of direction as you navigate— especially when it is relatively quiet and the winds are nearly calm, as they often are at night. Some examples of sounds might be traffic on a main highway or the pounding surf, and helpful smells might be a freshly plowed field or some stench from a polluted site.

ESTIMATING DIRECTIONS

Once you have determined a direction (such as from the GPS unit or the North Star), the "clock method" for estimating directions may be very useful to you (**Figure 4-20**). Just envision north (08) as being represented by noon on the clock face and each of the succeeding eleven hours as being 308 greater than the previous one. Of course, it is also easy to interpolate azimuth directions that fall between each of the hours.

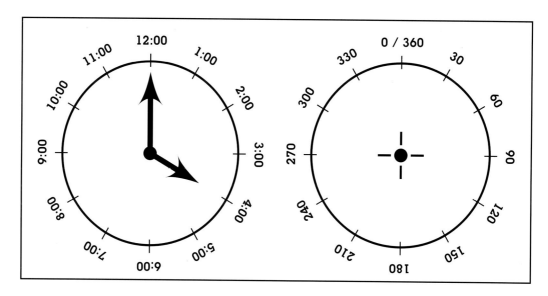

FIGURE 4-20

You might be surprised how accurately you can estimate directions using this method.

DETERMINING DISTANCE

Distances between any two stored positions (waypoints) are calculated and reported by the GPS receiver at the same time as direction through use of the navigation function. These distances can also be measured on the map using a piece of scrap paper and the graphic scales found in the margins (**Figure 4-21**).

GPS receivers report only straight line distances—not irregular ones. Thus, it is more practical to measure irregular distances, such as road or trail distances between locations, directly on the map.

To measure irregular distances along a winding road, or stream, a scrap of paper is again used. In doing so, the paper is placed along the edge of the road at the starting point. Tick marks are made both on the map and on the paper at the starting point and at the point where the road curves away from the edge of the paper. Keeping the second set of tick marks aligned, the paper is pivoted until its edge is again running along the same edge of the road. A third set of tick marks is made where the paper's edge and the route again separate. This process is repeated as many times as is necessary until the entire route has been marked off in this manner (**Figure 4-22**). Finally, the paper's edge is placed along the graphic scale to read the measured distance.

As was the case prior to the development of GPS technology, ground distances can be measured by an odometer, pacing (the average hiker takes 120 steps for every 100 meters), or elapsed time (using time-distance formulas). However, the GPS unit can keep track of your movement progress and report distances yet to be traveled in moving from your current position to the next waypoint (checkpoint) along your selected route.

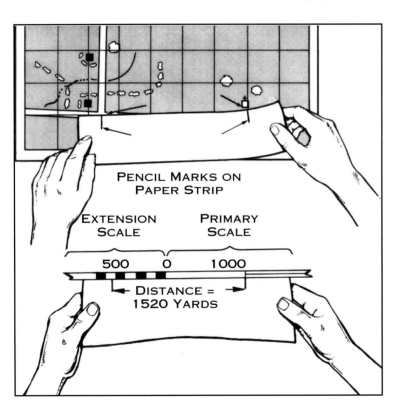

PENCIL MARKS ON
PAPER STRIP

EXTENSION
SCALE

PRIMARY
SCALE

500   0   1000

← DISTANCE = →
1520 YARDS

FIGURE 4-21

You are cautioned that ground distances calculated mathematically through use of the grid (as are those accomplished by a GPS receiver), as well as those actually measured over the flat surface of a map, fail to account for the discrepancies actually encountered out on the ground. Some of these discrepancies are caused by the unevenness of the terrain, which adds more surface distance to any horizontal calcula-

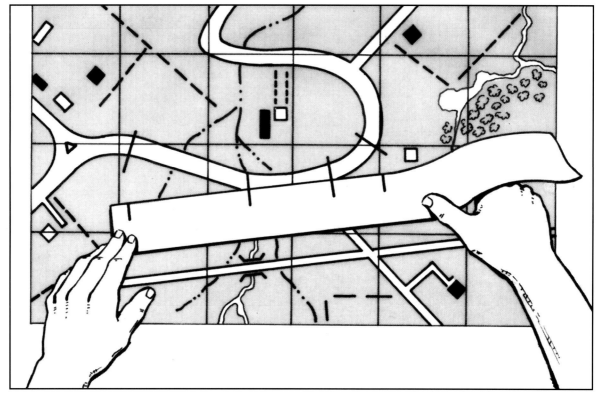

FIGURE 4-22

tions or measurements taken from the map. Other discrepancies are the result of the many slippages and small detours caused by slopes, soil types, mud, snow, the presence of rocks and boulders, wet areas, trees, depressions, small escarpments, and so forth. Therefore, there is a table of factors to be added to measured horizontal map distances before they are applied to your odometer readings, pace counts, or elapsed time distance calculations out on the ground.

FACTORS TO BE ADDED TO MEASURED HORIZONTAL MAP DISTANCES

Flat, scrub dessert or temperate terrain...................... add 10%

Rolling temperate wooded terrain .... add 20%

Any loose surface material or snow .... add 20%
(for wheel, track, and foot slippage)

Jungle or hilly temperate terrain .... add 30%

Sandy desert .... add 30%

Mountainous terrain .... add 40%

Source: U.S. Army Research Institute.

IN SUMMARY

We said earlier that in order to get there from here, you must know which way to go and how far. You should now be able to determine both.

This chapter explained how to integrate use of the magnetic compass with GPS equipment and a map while navigating over the land. This included how to orient the map using a compass and terrain referencing and how to determine directions with a compass, the GPS receiver, and several field expedient techniques. It also reviewed techniques for using the GPS unit to calculate both directions and distances among various positions saved as waypoints. And, finally, it reviewed the technique used to measure distances on a map and how to factor those measurements for a more accurate application out on the ground.

360° PHOTOGRAPH BY AGNES LIPSCOMB

EMPLOYING GPS

Now that you have acquired map and compass using skills and have become familiar with the many applications and advantages of using today's highly reliable GPS equipment, it's time to learn how to integrate and routinely employ them as you move from place to place (Figures 5-1, 5-2, & 5-3).

BACKGROUND

Since 1989, when the Magellan Systems Corporation introduced the world's first hand-held GPS receiver, a host of manufacturers and models have entered the arena. By way of illustration, a recent <u>GPS World</u> "Receiver Survey" listed a total of 58 manufacturers producing all types of GPS equipment. From this group of 58, there were 15 reported to be marketing a total of 28 portable models specifically designed for use in land navigation and selling for $2000 or less. Of these 28 models, 18 carried price tags of under $1000, and 3 were selling for less than $500. However, now that prices are about to drop into the $200 range, it is obvious they must flatten out.

With the frequent model changes being introduced at the present time, it is not practical to make unit-by-unit comparisons in a published book. That can best be left to the monthly periodicals. One such comparison of five hand-held GPS land-based receiver units was recently published in <u>Outside</u> magazine as part of its "Special Issue Buyer's Guide." No doubt, they will become regular features in a number of magazines.

In the one short year since the publication of this book, a host of improvements were made relative to the hand-held portable GPS navigational equipment now available to consumers. Among these improvements are a modest number of new features: some innovative graphic screen presentations, further reductions in size, weight, and price, increased battery life, and, not to be overlooked, the development of much more readable and understandable user manuals.

FIGURE 5-1
MAGELLAN GPS 2000

These changes prompted the publisher, Alexis USA, Inc., to take advantage of the need for a second printing by, instead, producing a second edition. This chapter on employing GPS has been completely revised; however, excepting a few minor corrections in the text of the other chapters related to the use of maps, compass, and so forth, the remainder of the book is unchanged.

Given that most hand-held GPS receiver units have similar features and characteristics and that the folks at Magellan have enthusiastically provided considerable amounts of technical assistance and information requested for this project, models from their latest inventory have been used to illustrate the various features being discussed. When one of their products does not adequately explain a particular teaching point, another manufacturer's device has been used. This in no way limits the usefulness of the book to those owning a particular brand or model. Our purpose is always to meet the needs of all who wish to apply the advantages of this new technology to their navigation requirements on land.

Finally, a quick look into the near future tells us that unit price reductions are not the only aspect of GPS marketing that is about to slow down. The development of new features and functions will do so as well. The improvements to come next in this portable equipment will be less about what the units can do and more about how well they can do it. These changes will most likely involve better refined equipment characteristics such as increased speed, less drain on the batteries, continued size and weight reductions, improved graphics displays, and more advanced antenna systems.

THE SITUATION

You have read the first four chapters on general navigation skills and now wish to proceed step-by-step with your new GPS receiver in hand in order to achieve maximum advantage from this powerful new tool. Because the majority of map illustrations are taken from the Easton, PA, area, you are to assume that you are located there for the purposes of your practice in initializing and setting up your unit.

When you are finished with this chapter, you will understand and be able to effectively employ any function and feature found on most GPS receiver units now available. Furthermore, you will know how to program these sets to better meet your individual needs and preferences.

Finally, in order to better organize the presentation and make it more understandable, the subject matter has been divided into the following topics: (1) receiver unit characteristics, (2) initialization and set-up procedures, (3) the four GPS navigation functions and their related features, and (4) other auxiliary features and unit accessories.

GPS RECEIVER UNIT CHARACTERISTICS

Today's portable receiver units are tough, reliable, resistant to dust and water, and they weigh less than a pound. Most will fit into the palm of your hand. In addition, they operate without difficulty through a wide range of temperatures as they report crucial position and other related navigation information clearly through a combination of words and numbers as well as through innovative, easily understood graphic presentations.

As surprising as it seems, unit accuracy is not related to price. Instead, price is more likely to translate into the numbers and types of features and capabilities included with the unit. More expensive receivers are often designed to accept external power sources and active antenna systems, hold greater numbers of positions as landmarks (waypoints) and preplanned routes and route segments in memory, and accept various electronic and communication interfaces. They are also more likely to include a capability for differential GPS (DGPS) (Appendix B), perform position and speed averaging computations, and accomplish more sophisticated location projection and triangulation functions, both of which are explained later in the chapter under the section on auxiliary features. Keen competition, however, is forcing manufacturers to include more of these features on their most economical models.

Two fundamental characteristics common to all portable GPS receivers are their rather heavy draw on

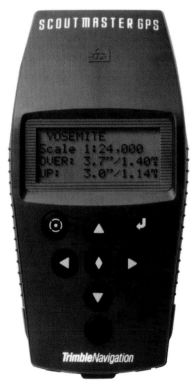

FIGURE 5-2
TRIMBLE SCOUTMASTER GPS

battery power and the need for proper set placement and antenna orientation. You should frequently monitor battery power and always carry fresh replacements. Rechargeable batteries are accepted by some models; however, you are warned that they may drop to an unusable level very rapidly and, unless they are replaced immediately when this occurs, you may lose all stored information. These data losses will include your present position, all saved locations, routes set, and so forth. Subsequent to any loss of power, it may take the unit 15 minutes or more to obtain the next position fix because it must first collect and store a new almanac from an available satellite before it can calculate its position. A GPS almanac is the data held in the receiver unit regarding the correct date, time, and the location and health of each satellite in the NAVSTAR constellation.

For maximum performance, you must place the unit so that it has an unobstructed "view" of as much sky as possible and with its antenna oriented directly upward. Some units are capable of tracking up to 8 or 12 satellites simultaneously, which speeds up the automatic process of switching to another satellite each time the signal from one of the four being used in continuous position calculations is lost or when the quality of the geometry being used to make the calculations can be significantly improved. In other words, your receiver should not be shielded by vegetation, structures, or the terrain. Locations in heavily forested areas, steep and narrow canyons, and inside or up against the outer walls of buildings will generally preclude your obtaining either a good reading or any reading at all. When this occurs, the best corrective

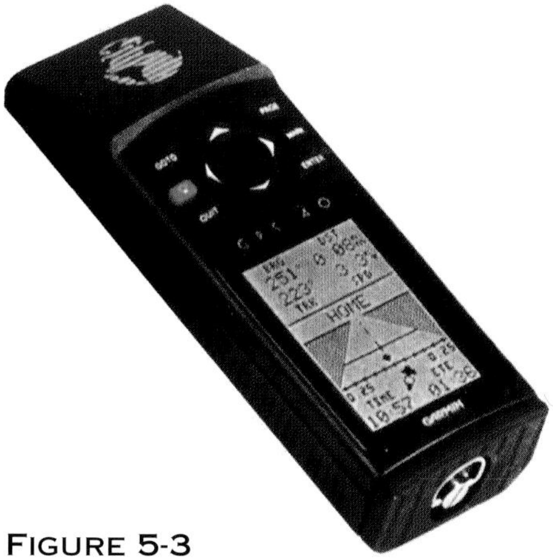

FIGURE 5-3
GARMIN GPS 40

action to take is to move with your unit to a more favorable location and try again.

Remarkably, the strength of the signals being received from the various satellites has little impact upon the accuracy of any position location being reported. As long as the information has been received by the unit, it can be used. However, poor signal quality (SQ) messages warn that a particular satellite's incoming data may soon be lost. Unless there are ample numbers of signals coming in from alternative satellites (a total of four for a 3-dimensional fix and three for a 2-dimensional fix), you may not be able to quickly obtain another reading as the result of various obstructions situated close by. On the other hand, the geometric angles resulting from the locations of the satellites being used by the receiver to calculate its position can drastically affect its accuracy. Whenever the geometric quality (GQ) is not acceptable, most GPS units will provide an on-screen warning to caution the user about the questionable accuracy of the calculation.

In addition to the warnings about low battery power, weak SQs, and poor GQs, most GPS units employ a variety of logical on-screen messages, icons, and graphics to convey various warnings and other vital pieces of information. Sometimes, they even use audible alarms. Here is a list of the most common examples:

- position information being reported is old (generally in excess of 10 seconds)

- position reported is based upon a 2-dimensional calculation (only 3 satellite signals)

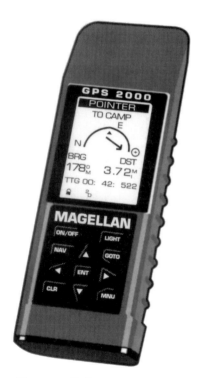

FIGURE 5-4
MAGELLAN GPS 2000

- screen backlighting is turned on and consuming additional battery power

- end leg landmark or final route objective has been reached

- you are off course and the cross track error (XTE) exceeds margin set for the warning

- up, down, left, right arrows appear to indicate that these keys can be pressed to select menu items, access further information, or to input data

- battery power is low (replace soon)

- batteries must be replaced or data will be lost (re-place immediately)

- external power has been lost (only when using an external power source)

The last general characteristic we will examine is the commonality evolving in unit keypad design (Figures 5-4 & 5-5). The keypad is obviously the means by which you interact with the unit. Most function menus are accessed by pressing various labeled function keys (e.g., setup, location, landmark (waypoint), route, and navigation) or some other menu keys. Four directional arrow keys are also usually included. They are used to scroll through lists of menu items or stored data, access additional information reports or graphic screens, or to "type in" alpha and/or numeric data being called for and to move the cursor about the screen. Any instructions or data you either select or type in are inputted by pressing the Enter key. Conversely,

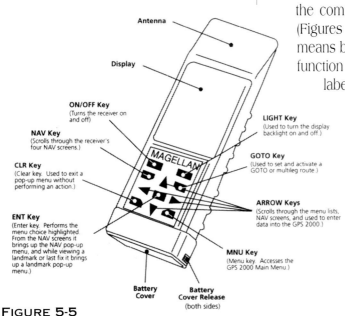

Antenna

Display

ON/OFF Key
(Turns the receiver on and off.)

NAV Key
(Scrolls through the receiver's four NAV screens.)

CLR Key
(Clear key. Used to exit a pop-up menu without performing an action.)

ENT Key
(Enter key. Performs the menu choice highlighted. From the NAV screens it brings up the NAV pop-up menu, and while viewing a landmark or last fix it brings up a landmark pop-up menu.)

LIGHT Key
(Used to turn the display backlight on and off.)

GOTO Key
(Used to set and activate a GOTO or multileg route.)

ARROW Keys
(Scrolls through the menu lists, NAV screens, and used to enter data into the GPS 2000.)

MNU Key
(Menu key. Accesses the GPS 2000 Main Menu.)

Battery Cover

Battery Cover Release
(both sides)

FIGURE 5-5
MAGELLAN GPS 2000

various instructions and data are generally deleted by pressing the Clear key. There is a Power key to turn on the set and often a Light key to illuminate the screen for nighttime use. On the other hand, some manufacturers have included a keypad similar to that of a push button telephone because they obviously believe it makes the entry of alpha-numeric data easier than use of the up and down arrow keys to scroll through numbers and the alphabet (Figure 5-6).

UNIT INITIALIZATION & SETUP

There are a few steps you're required to take before your GPS receiver is ready to find your position and lead you across the wilderness. You must first initialize it and then tailor its setup specifications to meet your particular preferences. After insuring that your unit has a fresh set of batteries, you are ready to begin.

INITIALIZE

In order to save time in obtaining your first position report, it's necessary to give your set some idea as to where in the world it has been awakened. This will not be necessary when the last location it computed was within 300 miles/482 kilometers of your present position (PP).

FIGURE 5-6
EAGLE ACCUNAV SPORT GPS

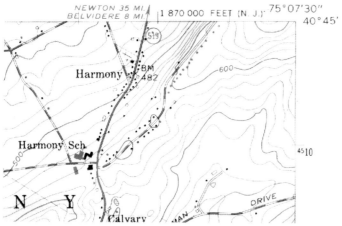

EASTON QUADRANGLE
NEW JERSEY—PENNSYLVANIA
7.5 MINUTE SERIES (TOPOGRAPHIC)

FIGURE 5-7

Your initial position can be entered as part of the unit's "setup function" (see below) in LAT/LON or any other coordinate system (including UTM or the MGRS format of UTM), depending upon what coordinate systems are carried in its software and which has been selected by you. Even when you plan to use a coordinate system better suited to land navigation, it is probably easiest to enter your initial position in LAT/LON coordinates. This information can often be taken directly from any corner of a large or intermediate scale map of the area in which you are navigating. For example, Figure 5-7 shows the upper right corner of the Easton, PA, NJ, USA 1:24,000-scale USGS 7.5 -minute Series topographic map quadrangle. Since we are planning to navigate using this map, you should input<40°45' N Lat, ENTER, 075°07.50' W Lon, ENTER, 720 (feet) ENTER> as your initial position.

SET-UP

Now, let's review the specific parameter selections you can make in tailoring your unit to best meet your needs relative to the various situations you may encounter while navigating with a map of Easton. If you own a GPS unit, you are encouraged to practice setting up each of these common options on your set. At the same time, you are encouraged to have you owner's manual handy, just in case you need to refer to it.

Coordinate system setup options available on various GPS devices offer you a number of choices for your position report displays. For example, as a minimum, most units include LAT/LON and UTM grid options. Incidentally, within both the Magellan Trailblazer and Trimble Scout model groups there are sets available that also offer the MGRS-UTM 1, 2, or 3 options. Furthermore, the Trimble ScoutMaster includes an innovative "over & up" coordinate reporting strategy (Figure 5-8). This particular option allows you to work on a map displaying no coordinate lines by simply entering the LAT/LON of the south-east (lower right) corner reference point and the scale of the map. Through its position reports, the set will then tell you how to measure the distance in inches and tenths of inches (or centimeters) over from the left hand margin of the map and up from the bottom margin to your position.

There is no question that this strategy works, but it may not always be convenient to stop and slap a ruler on the map in order to take the necessary measurements. In those cases, it would have been preferable to use a map prepared with a standard grid coordinate system drawn boldly across its face, even if, beforehand, you had to prepare the map yourself. (Procedures for preparing your map for use with GPS are described in Chapter 3). Regardless, Trimble is to be commended for its creative effort to successfully overcome the problem resulting from the dearth of street and highway maps currently available for use with GPS.

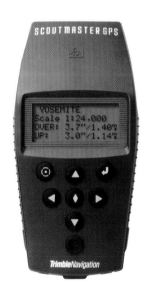

FIGURE 5-8
"OVER & UP" POSITION SCREEN
ON TRIMBLE SCOUTMASTER GPS

See page 137 for a more specific discussion of a proposed solution to this problem.

The coordinate system you selected during unit setup should, of course, match the one found on your map. However, if both LAT/LON and UTM are included (as is the case for the EASTON map), it is suggested that you opt for the UTM because its perpendicular grid better accommodates navigation on land, as explained in Chapter 3. Most GPS units have LAT/LON set as the default because your initial position is most easily entered from the true geographic coordinates shown in the corners of most maps. These GPS units allow you to set the angular LAT/LON position reports to be displayed either in degrees and minutes (including decimals to hundredths) or in degrees, minutes, and seconds. On the other hand, UTM Grid coordinates are reported as 6-digit easting and 7-digit northing values (to within 1 meter). And, when reporting in the MGRS format of the UTM coordinate grid, receivers can usually be set to display positions to the nearest 10,000 meters (2-digits), 1,000 meters (4-digits), 100 meters (6-digits), or 10 meters (8-digits). Since we must contend with the inaccuracies of the government's SA (selective availability) program and the fact that being within 100 meters is close enough for most purposes when navigating on foot, it is suggested that you select the readout to within 100 meters. When traveling on roads, you may wish to change the setting to within 1000 meters (1 kilometer), and, when driving the interstate highways at higher speeds, you may prefer a setting to within 10,000 meters (10 kilometers). Don't forget, estimates are easy to make down to tenths of a grid square.

Although LAT/LON is the default on most GPS receivers, our large-scale map of the Easton area includes the UTM Grid; therefore, it is suggested that you select either UTM or its easier to read MGRS version for your position reports. It is important to note that some units, such as the Magellan Trailblazer XL, will report positions in a second coordinate system by simply pressing the LOCATION key a second time.

PROPOSED SOLUTION TO THE MAP GRID PROBLEM:

Hopefully, our map publishing houses will soon address this serious shortcoming in their products. It is quite necessary that they do so if we are to gain maximum benefit from the speed and accuracy offered by GPS supported navigation techniques. Obviously, this improvement is very important to the growing body of GPS consumers and, in fact, to all of us who might require the emergency services of various public safety agencies. Each map maker and GPS manufacturer is cautioned against adopting its own individual location grid for use on maps because this will do a great disservice to consumers and fracture or compartmentalize both map and GPS markets to everyone's disadvantage. We need a standard grid common to all maps and all GPS receivers—one that has the following characteristics:

1) of maximum utility for use on land

2) relatively simple and easy to comprehend and use by the general public

3) does not interfere with the use of maps by non-GPS users

4) quick and inexpensive for map producers to re-search and apply to new issues of the maps they currently have in publication

5) cost-effective for incorporation into the software presently in use by GPS receiver units

It seems that the MGRS format of the UTM grid is best suited for this purpose. 1) The perpendicular UTM grid was specifically designed for use on land, while the angular nature of Lat/Lon, for example, provides no consistency of spacing as it has no perpendicular grid. 2) Whereas the regular UTM format calls for position reports consisting of 13 digits (6-easting & 7-northing), the MGRS 6-digit format of the UTM is the easiest to understand and use by the general public. 3) Crisp, sharp grid lines with bold one- or two-digit labels should assist all map users. 4) UTM grid information is available on all large and intermediate scale USGS maps, which serve as base maps for all cartography houses and are available for use by everyone in the public domain. And, 5) MGRS-UTM software has already been developed and is being used in some models by nearly every GPS manufacturer. To illustrate the ease with which this grid system can be used, consider this analogy using a common telephone number (610-252-1234):

AREA CODE	EXCHANGE	SPECIFIC TELEPHONE
(610)	252	1234
CENT EAST PA	EASTON	RESIDENCE

Now look at the parallel concept apparent when using the MGRS format of a UTM grid coordinate:

REGION	LOCALITY	POSITION		
18T	VA	8 82 825	0 05 052	(10km) (1km) (100m)
EAST NY /PA	EASTON AREA	MAP COORDINATES		
6° X 8° GRID ZONE DESIGNATION	100,000 M SQ. ID.	LOCATION TO ( )		

If such terms as Grid Zone Designation, 100,000-Meter Square Identification, and Specific Map's Coordinates are too imposing or sound too military for the general public, then simply substitute the terms "REGION," "LOCALITY," and "POSITION" in their places.(Refer to Chapter 3, Fig. 3-23, & App. A for details.)

SET-UP (CONT.)

The second setup feature we will discuss is the north reference option. It simply requires you to choose between magnetic north and true north. When operating with a map and compass, it is wise to instruct your receiver unit to report directions that coincide with your compass (magnetic north). Automatic magnetic declination corrections for all areas of the world have been built into the unit's software.

The time display setup parameter provides two options: (1) universal time coordinated (UT), until recently known as Greenwich Mean Time (GMT), and (2) local time. In most cases, you can choose to have local time reported on a 24-hour basis or in terms of AM and PM.

The map datum setup menu allows you to scroll through a listing of the various datums available on the set. Your choice should match the datum used for drawing the map sheet you are using, which is generally reported in the legend or somewhere else in the margin. A map datum refers to the theoretical mathematical form of the earth's sea level surface upon which the map maker has based his cartographic rendering. Position reports will differ somewhat from one datum to another. Errors of up to 600 meters can occur when an incorrect map datum is selected on your GPS receiver for the map at hand. Many units use as a default WGS-84, which is a good guess when you aren't certain which one to select. However, older U.S. DMA and most USGS topographic map quadrangles currently in use are based upon NAD-27 (North American Datum of 1927), which should be selected on your set. Some other datums from around the world that are commonly found on receivers include Europe, Alaska, OHAWA (Old Hawaiian), GRB 36 (Ordnance Survey of Great Britain), RT-90 (Sweden), KKJ (Finland), Tokyo, WGS-72, Australia Geodedic-84, and SAM-56 & SAM-69 (South American). While working with the EASTON, PA, NJ, USA, 1:24,000-scale (7.5'-Series) USGS map sheet, the NAD-27 datum is correctly used.

> **Helpful Hint: When you have no idea as to which datum was used in developing the map or when the appropriate choice is not available in the unit, go to a location about which you are certain both on the map and on the ground (e.g., a bridge or road junction) and then experiment by obtaining position fixes using different datums. Select the one that gives you the most accurate GPS reading, as compared with the map.**

The position elevation mode setup option in some GPS receivers offers three different choices: 2-D, 3-D, and automatic. When given a choice, the automatic mode should always be selected. If not, select 3-D. Two-dimensional position reports should be used on land only when the unit cannot locate and receive signals from a minimum of four satellites upon which to base its calculations. When set in the automatic mode, the position reported on the screen will be 3-D unless it includes an icon or report telling you that the unit is operating in a 2-D mode. A 2-D position report can be far less accurate on land because the elevation of the antenna is factored in when the set makes a horizontal position calculation. Now try setting your unit to the automatic mode.

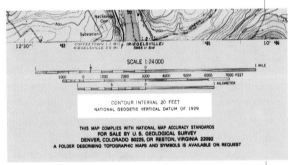

FIGURE 5-9

The distance/speed setup options available on most sets allow you to choose navigational reports in terms of statute miles/MPH, nautical miles/knots, or kilometers/KPH. While navigating when using topographic or any other types of maps showing UTM grid coordinates, it is suggested that you set the unit to report distances and speeds in metric units because the grid lines are generally either 1 or 10 kilometers apart. Incidentally, the diagonal distance across a 1 kilometer grid square is 1.4 kilometers. Also, one tenth of a kilometer is obviously 100 meters, which is the distance from the goal line on one end of a football field to the goal posts at the other. Also, 1 kilometer equals .6 mile and 1 mile equals 1.6 kilometers, approximately. It is plain to see that it isn't at all difficult to begin to think in terms of metric units.

The elevation units setup option offers a choice between feet and meters. Again, you should check the units found on the map you are using. The Easton area map sheet prompts us to set the unit's elevation parameter in feet (Figure 5-9).

Screen backlighting brightness can usually be controlled as a unit setup option. The feature is, however, generally turned on and off through a keystroke.

Screen contrast settings for most units' displays are also frequently controlled as a unit setup option. Ambient temperature has an effect on the display, thus some adjustments may be required as these conditions change. Just as for backlighting brightness, most units show a graphic scale to help you select your best contrast setting. You are now encouraged to stop and practice making these adjustments.

Foreign language options are available on some units.

Various timing adjustment options allow you to determine how frequently automatic position fixes will be taken and the time interval can be set between automatic saves in the unit's "last location buffer." Finally, there are models available that also allow you to set the time span you wish to use in conjunction with an automatic shut-off feature. This, of course, helps the absent minded navigator to conserve precious battery power.

FOUR GPS NAVIGA-TION FUNCTIONS & RELATED FEATURES

Next, we will review in some detail the capabilities and applications of the four primary navigational functions available on virtually all GPS receiver units. They are (1) reporting position, (2) storing landmark (way-point) names and locations, (3) planning and storing single or multi-segment route information, and (4) navigation (following the selected route).

POSITION FUNCTION

By simply powering up most GPS units, they will report positions in the pre-selected map coordinates. Figure 5-10 shows three GPS unit screens reporting the same position in (a) LAT/LON, (b) UTM, and (c) MGRS format UTM coordinates.

As another example of what can be done, Trimble Navigation Limited and Thomas Brothers Maps have teamed up to devise a new positioning grid and produce a small scale national road atlas, entitled The Trimble Atlas, with map sheets displaying this rectangular grid. It is composed of regional maps (scale of 1:1,900,800 or 1 inch to 30 miles) covering the contiguous 48 states and a scattering of larger scale area maps (scale of 1:316,800 or 1 inch to 5 miles), and it can be used by motorists in conjunction with Trimble's ScoutMaster GPS receiver.

FIGURE 5-10
POSITION REPORT - LAT/LON
POSITION REPORT - UTM GRID
POSITION REPORT - MGRS

When using this coordinate reporting feature in conjunction with the Trimble Atlas, the position reports appearing on the ScoutMaster's screen will tell you first on which page of the atlas you are located and then place you within a 49.5 x 38.5 mile rectangular grid area on that particular regional map. Furthermore, through use of an on-screen icon, it will more precisely position you within that grid space in one of nine sub-areas (upper left, center left, lower left, and so forth). See Figure 5-11.

One major disadvantage of this strategy is that the scales are really quite small, thereby eliminating many important details and most of the roads from the maps' portrayal. Also, few of the more localized areas included within each of the rather sizable regions are portrayed on the larger scale area maps. For example, the Atlas contains a total of only seven Area Maps within Region 6 (Pacific Southwest), while Region 9 (Southeast) has only four and Region 2 (Mountain) has none. And, finally, another major disadvantage is the fact that the grid coordinate system has been copyrighted by its publisher, legally limiting the number and types of maps that can display it and seriously curtailing its overall usefulness to both consumers and producers of GPS equipment.

In summary, until map producers begin the practice of showing the lines of a standard grid coordinate system on their products, the most viable solution to the map selection problem is to follow the instructions discussed earlier. They told you to select the map you find most useful when navigating in a particular area for

FIGURE 5-11
GPS SCOUT POSITION SCREEN
FOR USE WITH TRIMBLE ATLAS

your intended purpose and then to prepare it for use in conjunction with GPS. Then, after consulting Chapter 3, Appendix A (MGRS-UTM Reference Map of the U.S.) and, finally, the USGS base map sheets for the same area covered by the map you selected, carefully copy the UTM grid lines on your map and include the "shorthand" MGRS-UTM labels (1 or 2 large numerals only) for each line. This can be done for any scale map you might select, including the strip trip maps you get from your motoring club. If you plan to navigate using large scale USGS topographic maps not displaying the grid, you need only connect the blue UTM grid tick marks found in opposite margins outside the neatline of the map and add larger labels that are easier to read.

The time you spend in preparing a good map for use in conjunction with a GPS receiver will pay great dividends in terms of the enjoyment, speed, and accuracy you are able to derive from use of your equipment.

LANDMARK (WAYPOINT) FUNCTION

As they continue to design navigation receivers for use specifically on land, GPS manufacturers are beginning to overcome their nautical roots and drop some unaccustomed terms such as "waypoint" and "man overboard" and to talk about "landmarks," "checkpoints," or just plain "locations." The purpose of the landmark function on any GPS receiver is to either automatically or manually store in the unit's memory the position information for various designated locations.

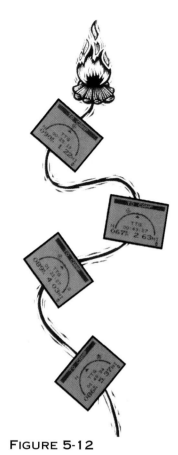

FIGURE 5-12

There are a variety of ways this information can be categorized and held, but the purpose is always the same. It will be available whenever you wish to use it to establish a route and find your way.

Whenever you take a position fix, the information defining it—including its horizontal coordinates; vertical position; time/date stamp; and, perhaps, even the satellites used to compute it, the strengths of their signals, and the geometric quality of the fix—can be automatically stored in a library of position saves. Furthermore, they can either be chronologically numbered, beginning with 001, or named by you so they can be more easily remembered and used later on. Finally, map locations you have not visited, yet wish to store in memory, can also be manually entered into your landmark library and, again, each can be automatically numbered or named by you. And, of course, you can delete, edit, or change the names of any of these stored landmarks at any time, unless they are currently being used to help define part of a stored route.

In addition, most units automatically save your most recent fixes in what is often called a LAST LOCATION BUFFER. For example, up to 15 positions (or whatever is called for by the design of your unit), will be held, while each new save replaces the oldest one still being held. Magellan units also have a BREADCRUMB feature that saves with an "+LLC" designation in the LAST LOCATION BUFFER every position you wish to keep by quickly pressing ENTER, ENTER, ENTER. Whenever this feature is switched on during SETUP and when your path has a number of key turns or junctions, you can save a BREADCRUMB landmark at each deci-sion point. At the end of your trek, you can create a BACKTRACK route using all the key locations from the trail of "bread crumbs" you "dropped" along the way into you unit's memory.

Yet another feature commonly found within the landmark function on most GPS receivers is that you can scroll through the list of landmarks in the library by numerical, chronological, or alphabetical order (some-times even by proximity order) to review, edit, or delete any of the information being held.

ROUTE FUNCTION

FIGURE 5-13
MAGELLAN POINTER SCREEN

This function includes features that allow you to develop, store, and select a number of routing options in preparation for your navigating over them. For example, using landmarks already stored in your library to define the various route segments (i.e., from camp to bridge to RJ 1 [road junction], etc.), you can create and store in the unit's memory a multi-segmented route. Some devices will allow you to plan several routes in advance and activate the one you wish at any time. In addition, by using menu choices and a single keystroke (normally ENTER), you can reverse the preplanned route for a return trip, activate a BREADCRUMB route, call-up a series of positions for backtracking from the RECENT LOCATION BUFFER, or activate a quick, simple GOTO route from your PP (present postion) to any saved landmark.

NAVIGATION FUNCTION

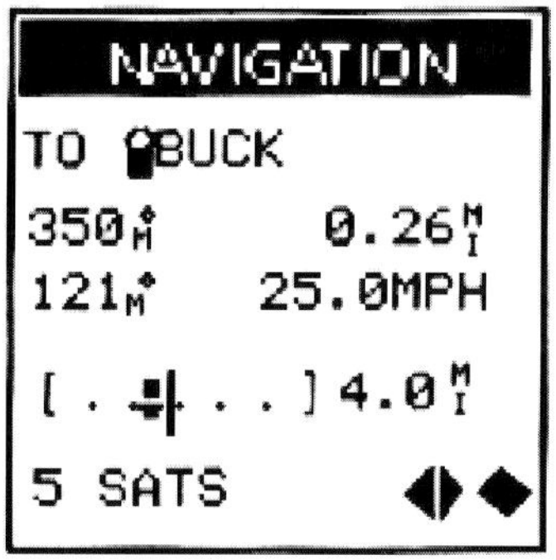

FIGURE 5-14
MAGELLAN NAVIGATION SCREEN

Use of the navigation function enables your GPS unit to "steer" you along a prescribed route; be it over the leg of a preplanned multi-segmented route, a quick GOTO movement to any selected landmark, your electronic BREADCRUMB trail, or the same steps back to a position captured in the LAST LOCATION BUFFER. Figure 5-12 illustrates what is meant by having a GPS unit "steer" you along a route. In this case, you see Magellan's unique "pointer screen" leading a navigator to camp. Figure 5-13 provides an enlarged image of a similar screen showing that the campfire is on an azimuth of 86° magnetic, 5.37 miles away, and, provided you maintain your current rate of progress, the time to go (TTG) before warming yourself at the fire is about 1 hour and 49 minutes. The "X" inside the circle represents the direction of the camp relative to North (N) and to the direction of travel, which is represented by the arrow pointing toward the top of the screen. You are reminded, however, that the GPS unit is not a compass and if you stop to talk and turn your body to face a fellow hiker, the arrow pointing in your direction of travel will maintain its position on the screen. The set can only orient itself while you are moving and as long as you don't change the orientation of the set.

The newest GPS units feature a broad array of navigation function screens to assist you in keeping to your route and recognizing when you have reached either an intermediate objective or the final destination. To provide further assistance, some units sound audible warning alarms to tell you when you have strayed a preset distance from a designated azimuth bearing or when you have reached either an important decision point along the way or the final destination.

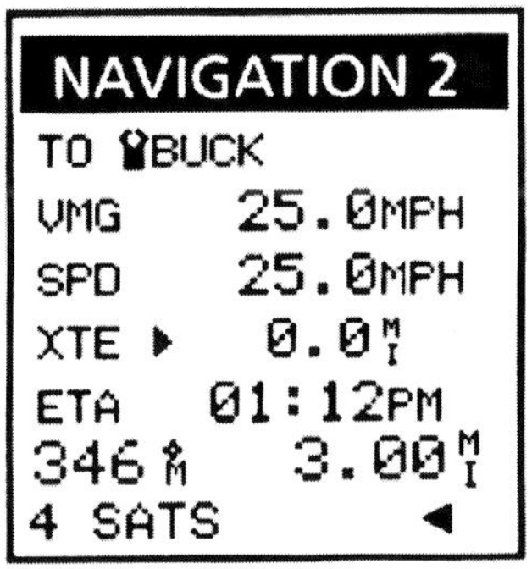

We have already examined Magellan's unique "pointer screen" (Figure 5-13); so let's take a look at several of the others. Most will be illustrated through use of the screens available on the Trailblazer XL model, as they are typical for our discussion.

Provided you are following a defined route, the first navigation screen (Figure 5-14) that appears when using this function, reports the name of the landmark toward which you are navigating on the top line of the screen. In this case, it is named "BUCK." The next line displays the azimuth (bearing) and distance from your PP to that landmark. The third indicates your actual heading and the speed you're moving. Next, comes the course deviation indicator (CDI), which graphically shows your cross-track error (XTE) and which way (left or right) and how far you have drifted off-course. Each dot on the scale represents 1/4 of the unit of measure preset by you for use on this CDI display.

The second navigation screen (Figure 5-15), which you may encounter on some units, usually includes a wealth of additional information. For example, this one is available on Magellan's Trailblazer series; and, in addition to the name of the destination landmark, azimuth direction (bearing) and distance to the destination, and the number of satellites used to determine your PP in order to accomplish these calculations, the screen allows you four lines of additional data that you can

FIGURE 5-15
MAGELLAN NAVIGATION 2 SCREEN

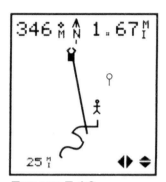

FIGURE 5-16
MAGELLAN PLOTTER SCREEN

tailor to meet your own specifications. As you can see from this figure, the seafaring blood has not yet been completely purged from the programmers of these units. Land navigators most frequently refer to bearings, headings, and steerings as azimuth directions. Oh well, we can certainly allow them to hang onto a bit of tradition while we enhance our vocabulary.

The information items from which you can choose to set up lines 1-4 of this second navigation screen are as follows:

Velocity Made Good (VMG)	is the portion of the velocity (speed) associated with moving closer to the destination as you are traveling on a path that will miss that objective. Another way to look at it is how quickly you are getting closer to the objective. If you are not on track and will miss your objective, at some point your VMG will obviously become a negative value.
Speed of Advance (SOA)	is similar to VMG in that it measures the part of your speed that is in the direction of the destination. But in this case, you have drifted off course, made the correction, and are now heading directly at the objective. SOA is then concerned with the comparative speed with which you are moving toward the objective relative to your theoretical position on the original track that you had planned to negotiate.
Speed (SPD)	is the velocity at which you are moving in respect to the ground around you.
Heading (HDG)	is the direction of your actual progress expressed as an angular azimuth or bearing.
Time to Go (TTG)	is the amount of time it will take to arrive at the next end of leg objective, based upon your previous progress.
Estimated Time of Arrival (ETA)	is TTG added to the current time.
Distance Made Good (DMG)	is the amount of linear progress made in the direction of

the objective since departing from the starting point. It is similar in concept to VMG and SOA.

Cross Track Error (XTE) is the distance you have drifted to the left or right of the planned directional path.

Steering (STE) is the correction in the heading needed to return to course. It is expressed as a new azimuth.

The plotter screen (Figure 5-16) will be examined next. It is found on many GPS receiver units and provides you with a track history of your movement, as well as the bearing and distance to the destination from your PP, as indicated at the top of the screen.

In addition to Magellan's unique pointer screen (Figure 5-17), already discussed, they also offer a similar graphic on some units called the steering screen. The pointer screen shown in Figure 5-17 is nearly identical to the one shown in Figure 5- 13 with the exception that the destination icon (X inside a circle) has been replaced by a buck to represent the landmark (buck), obviously serving as the end of leg objective. To steer yourself to the objective, you simply keep your heading arrow aligned with your destination icon. If not, you will develop an XTE and reduce your VMG, SOA, and DMG; and then you'll be required to make a STE adjustment.

The steering screen (Figure 5-18) accomplishes basically the same purpose as the pointer screen by showing you two portions of magnetic compass scales you must keep aligned. Obviously, if you don't have your heading (bottom) compass scale aligned with your

FIGURE 5-17
MAGELLAN POINTER SCREEN

FIGURE 5-18
MAGELLAN STEERING SCREEN

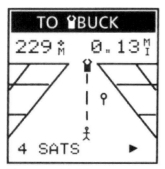

FIGURE 5-19
MAGELLAN ROAD SCREEN

planned direction of travel (top), you will again develop an XTE and reduce your VMG, SOA, and DMG; and you'll be required to make a STE adjustment.

Another graphic display that seems to be popular on GPS units is called the road screen (Magellan GPS = Figure 5-19, Garmin GPS = Figure 5-20, and Eagle GPS = Figure 5-21). In effect, the road screen is a pictorial CDI showing you the amount and direction of any XTE. The destination landmark is represented by the icon at the center top of the screen, the intended course of travel is shown as the dashed center line of the road, and the road's edges represent the CDI limits selected by you. Icons for any other nearby saved landmarks in close enough proximity also appear on the screen. If you are on course, the person icon is on the center line, and if you are to the left of course, for example, the icon moves to the left of the center line. The grid moves toward the bottom of the display in relation to your SOA. If SOA is negative or there is no movement, the grid remains stationary. Finally, the information at the top of the screen indicates that camp is on an azimuth of 229° magnetic and at a distance of .13 miles.

AUXILIARY FEATURES & ACCESSORIES

Next, we will take a brief look at some other auxiliary features and accessories available with portable GPS units.

FEATURES

Some of the special "bells and whistles" that can be found on various devices include the following:

FIGURE 5-20
GARMIN GPS 40

FIGURE 5-21
EAGLE ACCUNAV SPORT GPS

- multiple keystroke unlocking feature to prevent accidental turn on of the set

- selective erase features (landmark, route, or last location buffer, etc.)

- clear memory (erase all data held in memory and restore factory defaults in setup)

- set alarm parameters w/off & on capabilities (i.e., arrival at objective, CDI warnings, etc.)

- landmark sort (through stored landmarks by alpha, numerical, chronological, proximity)

- set frequency of position sampling and storing

- velocity averaging calculations

- triangulation projection to another position from two or more locations using azimuths

- landmark projection of another position from a location using azimuth and distance

- map projection of another position from a location using distances N & S from that location

- data uploading and downloading capabilities among various types of electronic interface equipment

- DGPS (differential GPS) capabilities (reception and utilization of signals from a nearby DGPS beacon)

- satellite status reports (positions, SQ, GQ, healthy, out of service, etc.)

- odometer screens to record overall distances and trip odometers that can be independently reset

- trip summary reports (where, when, how fast, averages, etc.)

- sun and moon data by location & date (i.e., time and azimuth of rise and set, waxing & waning moon, etc.)

- graphic grid overlay screen for various positions within selected coordinate system

- projected vehicular fuel consumption rates

ACCESSORIES

Various accessories can be purchased for GPS receiver units, as well. Here is a list of the most common items:

- carrying cases

- active extenal antennas that boost incoming signals w/various mounts

- lanyards

- receiver mounting brackets

- active extenal antenna jackets that fit over units with built-in antennas (e.g., Magellan GPS 2000 & Trimble ScoutMaster)

- external power units

- spare battery cassettes

- suction cup & magnetic vehicular mounts for detachable antennas

- various cables & couplings

- reference guides (user manuals)

IN SUMMARY

You should now have a thorough mastery of the information and concepts needed to initialize and setup your unit as well as apply the many functions and features available on today's products. Obviously, no single GPS model could possibly have all these capabilities, but the purposes of this chapter will be served if it has prepared you to know which features you prefer on your GPS receiver and if you are able to use any you may encounter. Although there is no claim that this chapter represents an exhaustive listing and review of all the features and accessories possibly available with portable hand-held GPS receiver units; every effort has made to make it substantially complete.

In conclusion, the best way to learn and retain navigation skills is to use them. So, gather up your receiver, map, and compass; and go out there and enjoy what GPS and nature have to offer.

FIGURE 5-22
THE MAGELLAN TRAILBLAZER

PHOTOGRAPH BY AGNES LIPSCOMB

GPS LAND NAVIGATION PRACTICE EXERCISE

(PLAN AND CONDUCT OR PARTICIPATE)

Whether you are the instructor or the instructed on the subject of navigating with GPS, you may wish to establish or participate in a short practical test to evaluate the effectiveness of your instruction or reinforce the wayfinding skills you are working to develop. This brief chapter will provide some suggestions for setting-up a short LN course and some helpful hints for negotiating one as a participant. As you might suspect, the GPS receiver will prove itself to be a valuable tool in either case.

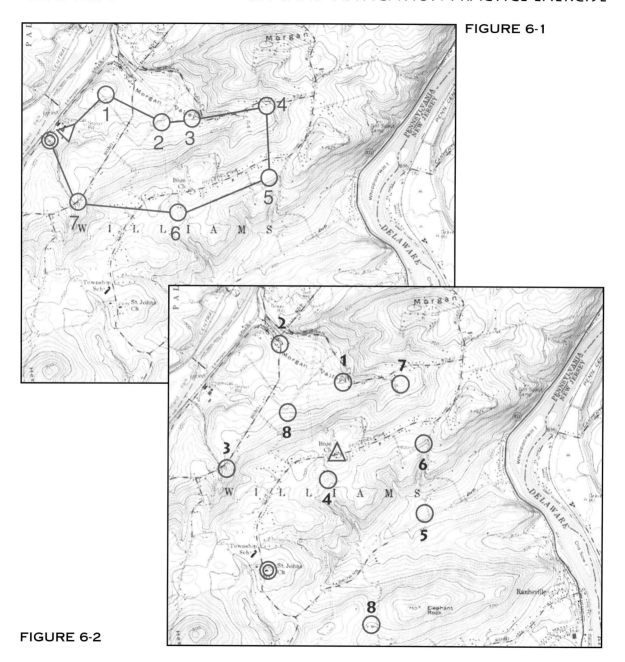

FIGURE 6-1

FIGURE 6-2

BACKGROUND

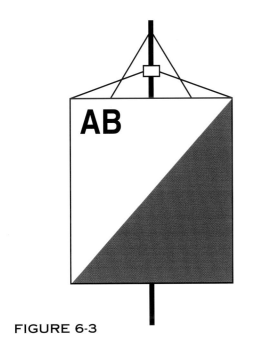

FIGURE 6-3

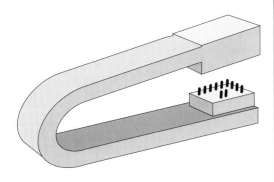

FIGURE 6-4

The design principles and techniques used to establish a simple orienteering course, more specifically a TRIM-orienteering course, provide an excellent basis for developing this type of practical, hands-on exercise. It will test not only the employment of GPS, but will encourage its appropriate utilization in conjunction with the terrain, a map, and compass.

Orienteering is both a competitive LN sport and a more leisurely recreational activity using natural outdoor areas and large-scale topographic maps. It may require participants to visit the established control points (waypoints) in numerical order over a specific course (Figure 6-1) or at random, as in TRIM-orienteering. The design of a course with several randomly placed control points is preferred for this activity in order to prevent "following" by the participants (Figure 6-2). Each group can be assigned a different set of two or three control points from common start and finish points without all participants having to wait lengthy periods for others to move on ahead and then follow the same route. This method encourages individualized route selection and better guarantees that each team will navigate on its own from place to place over the course.

Each control point is marked by a large orange and white "kite" hung clearly at a distinctive target feature (Figure 6-3) and its location is circled on the master control map (Figures 6-1 & 6-2). Every control point marker is coded (e.g., AB) and has a punch with a unique pin pattern to mark the score card as proof it was located by the team (Figure 6-4).

FIGURE 6-5

Very little equipment is needed to establish and operate a good quality course. Control markers and those unique punches can be purchased at various orienteering supply outlets and the necessary maps obtained from local outdoor stores, bookstores, the USGS, DMA, or any other appropriate governmental agency or commercial map source. Signs leading the participants off the highway into the test sight and master maps showing all control locations on the course should be fabricated and prominently displayed. Score cards can either be professionally printed or reproduced locally. And, you might consider issuing whistles to be blown in cases of emergency to facilitate finding and evacuating any injured personnel.

You may wish to contact the publishers of **Orienteering North America**, SM & L Berman Publishing Co., 23 Fayette Street, Cambridge, MA, USA, 02139-1111 (617-868-7416); The United States Orienteering Federation, P.O. Box 1444, Forest Park, GA, USA, 30051; or any other nation's Orienteering Federation for the names, addresses, and telephone numbers of reputable orienteering supply outlets.

Finally, distinctly marked parking and start and finish areas are helpful in better managing the various off-course activities associated with the operation of an LN test site. These activities include arriving and departing vehicular traffic, recording the locations of control points found on the master map(s) onto individual maps by the participants (Figure 6-5), and waiting for others to start and finish the course.

COURSE PLANNING

Before you start the process of planning an effective GPS LN skill test course, locate and obtain permission to use a natural outdoor site with sufficient area to accommodate the activity and the number of navigational teams you plan to exercise. You will require an area of at least several square kilometers in size. Generally, you should not locate more than two or three control points within a one square kilometer area, and you should plan to establish about twice as many control points as you have navigational teams operating at any one time on the course. Parks, other public lands, and military reservations often present excellent locations for this type of activity.

Next, you must obtain and study a large-scale map of the area to be used (1:50,000 or greater/1:25,000 preferred). This intensive map study should be done while walking over the terrain. Keep in mind the fact that you wish only to provide some practical experience in using the GPS while navigating, not brutally test the physical endurance or special mountain goatlike or swamp traversing skills of the participants. In other words, use terrain that is safe and relatively easy to negotiate. Also, be certain the entire course will be covered by a map that is fairly comprehensive and easy to read.

Now, it is time to select the specific locations of your numbered and coded control points. Their locations should be circled and numbered on the map and, later, a day or so before the course is to be used, the control markers accurately placed on the ground and coded with letters to provide proof of their being visited by the participants. The accurate placement of control points on the ground will be covered in the next section.

In terms of the course's length, it is reasonable to expect navigators on foot to cover 3,000 meters per hour during daylight or 1,000 meters per hour in darkness. Thus, nighttime courses can be much shorter, but care should be taken not to make control markers too difficult to find in the darkness. Use of lighted control markers is not suggested because they eliminate most of the challenge.

Control points should be placed on specific, identifiable terrain or other features in order to encourage use of map interpretation and terrain association skills by the participants. Some features that might be used for this purpose include a hill, the upper end of a small draw, lower end of a small spur, point along a small lake or pond shoreline, the point or tip of the arm of a swamp, a cabin, a location adjacent to a stream or road junction, and so forth. Obviously, these features should also be clearly portrayed on the map. Placement of control points on these types of features will not only help the participants navigate to them, but it will also assist you in placing the control markers more accurately on the ground.

Finally, control points should be placed in a way that discourages participants from wandering onto private land, damaging a fragile environment or habitat, and endangering themselves or property in any way.

COURSE SET-UP

First, you must clearly mark the parking area, master map viewing area, and your start and finish points and associated waiting areas. In addition, any signs and the master map(s) must be set out.

Next, you must accurately place all control point markers and punches at appropriate locations on the ground. Employ your GPS unit to generally locate your points and then use your map and terrain association skills to fine-tune the markers' specific placements. If a control point is shown on the map at the tip of a small spur, don't locate the marker in an adjacent draw out there on the ground. This will only cause your participants to lose confidence in your course and lead to "heated" discussions after the event.

Before employing the GPS unit to help you locate your control markers on the ground, it is suggested that you manually key in the coordinates and elevations of all control locations as waypoints. Then, you should proceed to each of their locations using your GPS receiver, compass and terrain association skills to correctly place the markers at each of the sites on the ground. It is always a good idea to have another knowledgeable person check your work.

CONDUCTING THE EXERCISE

In addition to a well planned and accurately prepared course, a good LN exercise requires a highly organized operation. Participants should be thoroughly briefed, quickly broken into teams, immediately sent through their paces, given a meaningful after-action review, and an adequate opportunity to have any questions answered.

The briefing should include general and specific safety considerations; the general layout of the course; what to look for and do at the assigned control points; any special cautions or helpful hints; a review of the schedule of events; and when and where to start, finish, wait, and participate in the critique and question and answer session at the end of the event.

Following the briefing, equipment should be checked and participants led to the master maps where they can mark the course control locations and numbers on their individual maps. Next, they should be led to the start waiting area where score cards, with the control numbers circled that the team is instructed to visit, will be distributed to each group. Then, teams must be sent out at regular timed intervals by the event starter.

After everyone has completed the course, all participants will engage in an after-action review to reinforce what has been learned. All questions from the participants should also be answered. It is crucial that the participants understand that their active sharing of experiences during the exercise is key to the effectiveness of this review activity.

HINTS FOR RUNNING A COURSE (AS A PARTICIPANT)

Actually, the first five chapters of this Guide present all the information, concepts, and skills needed to successfully complete a LN practice course or to meet any other LN challenge. It is assumed that by now you know how to employ a GPS receiver, interpret a map, and use a compass. However, a quick review of some helpful hints may serve as a good memory aid just prior to your running a course.

First, you are encouraged to mark on your map the locations of the various control points that make up the course with as much precision as possible. Further, you should take the time to manually enter the locations of your start and end points and any of the controls you are to visit as waypoints into your GPS unit's memory. You should also thoroughly analyze the map to select "functional" and easy-to-follow routes from the starting point to each of the controls you are to visit and on to the finish.

Remember to identify and use handrails, catching features, and navigational attack points where they will aid you in guiding your movements. You might also utilize other controls located on the course that are not required for your particular team's exercise, should they be helpful as guides. Finally, you are encouraged to take maximum advantage of any number of your GPS receiver's many functions and features and to fully integrate its application with your ability to use the terrain, your map, and a compass.

In conclusion, good land navigators exploit any advantage they may have available to them. This includes the fine capabilities of your GPS equipment and any cues offered by either the map or the real world.

IN SUMMARY

This chapter provided some suggestions for setting-up a short LN test course and some helpful hints for negotiating one as a participant. Briefly, you were encouraged to use some of the design principles and techniques to establish a simple orienteering course. Including the capabilities of the GPS receiver and your map and compass while using terrain association skills to establish or to run through a course exercise is an outstanding way to learn and practice the skills addressed in this chapter.

CHAPTER 7

LOOKING AHEAD (IMPACT OF GPS)

There is no question that GPS will have a profound affect upon everyone. In fact, it would not be at all surprising if our letterheads soon include our MGRS/UTM grid coordinates as well as our telephone and fax numbers, street addresses, and zip codes.

BACKGROUND

Some technological developments have rather limited impacts upon our daily lives, while others revolutionize the quality of life and the way we do things. Certainly, the development of the wheel, steam engine, transistor, and miniature computer chip all fall into that latter category. And, there is no question that the GPS is about to be added to that list.

A GLIMPSE INTO THE FUTURE

Most fundamentally, GPS is going to generate a vastly increased, broad-based interest in geography and maps. The spatial relationships among places and various human activities will steadily gain the attention of us all. GPS will inevitably contribute to our awareness that we all share the same planet, face the same challenges.

GPS obviously offers great utility to those who wish or need to navigate with precision, but it also holds as much potential for those who wish to apply and exploit its capabilities in nearly every other endeavor—academic, scientific, and commercial.

For example, new vehicular-tracking and reporting systems; pollution-tracking; port, airport, and highway traffic control applications; highly precise survey and underground infrastructure location inventory and recovery capabilities; and various geologic monitoring and experimentation strategies are just a few of the possibilities now being developed around GPS.

GPS equipment manufacturers and those developing and producing any number of ancillary electronic interface programs and devices will certainly find themselves in growth industries. Both paper and digital map producers (they each accommodate very distinct and viable markets) will experience increasing demands for their products because their maps will always need updating, and there will be new and different applications being continually developed that require maps to fill these highly specialized needs.

In addition, every commercial organization wishing to sell a product or service to the traveling public will wish to be included among those listed in the spatially-accessed inventory indexes now possible with the computer and GPS. No matter where you happen to be located in the world, there will soon be an electronic program that can automatically tell you where to eat, sleep, swim, shop for any number of products, picnic, hike to a view, and so forth. It will also tell you precisely how to get there.

Military operations,methods, and capabilities will certainly be as profoundly affected as those found in the civilian world. The spatial aspects of command, control, communications, fire control, intelligence, and logistical functions will ultimately be monitored and controlled through systems handling GPS position-and time-stamped data linked through various communications/computer networks that make complex and comprehensive real-time decisions possible on the battlefield. Larger units, weapons systems, and other functions will be operating with highly complex GPS-interface systems while small units, patrols, individual delivery trucks, and so forth will continue to function with stand-alone GPS units and paper maps.

Finally, civilian education and corporate and military training programs will increasingly emphasize geographic concepts, map reading ,and navigational skills.

Wayfinding will again be seen as a "survival skill," just as it waswhen ancient traders embarked for the Orient and when prairie schooners headed west to tame the American frontier.

IN SUMMARY

The title of an article found in the October 14, 1991, issue of **Aviation Week & Space Technology** best sums up the future of this revolutionary new space-age development. That title proclaimed, "Imagination Only Limit to Military and Commercial Applications for GPS". No one knows exactly where this new utility will lead, but it is certain that it will have a profound impact upon many aspects of our lives—how we travel, provide for the common defense and public safety, work, shop, and play.

APPENDIX A

Appendix A consists of the enclosed MGRS Reference Map of the continental United States and parts of Canada.

APPENDIX B

HOW DOES GPS WORK? (SOME TECHNICAL TALK FOR THE LAYMAN)

Most of us have a certain amount of curiosity about how new things work, especially when they are revolutionary and cost so many tax dollars. Without delving into the specifics of the complex physics, electronics, and mathematics involved, this is how GPS reports your position.

NAVSTAR GPS consists of three components: (1) space, (2) control, and (3) user segments. The space segment, consisting of 24 satellites orbiting the earth every 12 hours at an altitude of approximately 12,500 statute miles (about 20,200 kilometers), broadcasts the information needed by any user unit (receiver/computer) to calculate its position. Each satellite broadcasts highly accurate time information and generates a coded signal that is matched by the user unit to the time and coded signal being kept by its own built-in computer. The time discrepancy, measured in nanoseconds (1 billionth of a second) between them, is translated into distance calculations between the unit and the satellite. Since the user unit also contains an almanac of information telling it exactly where each satellite in the constellation is located at any given time, it can (in theory) determine precisely where the unit is located on the earth's surface in relation to any two satellites from which it receives signals.

Since theory and reality are not the same, information from a third satellite is needed to beef-up the accuracy of the fix. GPS satellites keep highly accurate time with very expensive atomic clocks, but individual user units do not have this kind of time-keeping capability. By knowing the orbital locations of all satellites at all times, these units can compare the information coming in from them and calculate the amount of time error existing on its own clock while continuously (but not perfectly) correcting it. Further, by using information from three satellites, the unit can further reduce its position calculation error by averaging it out. Of course, use of a fourth satellite allows the user unit to report its altitude as well as its horizontal position. The result is a highly accurate position fix every time the navigator calls for one.

The space segment was launched and is maintained by the U.S. Department of Defense (DoD) and represents a $10 billion investment. The responsibility for making small orbital adjustments of the satellites, maintaining the accuracy of code and timing signals, and upgrading the accuracy of almanac (satellite location) data being broadcast falls to the U.S. Air Force's control segment facility at Colorado Springs.

Selective Availability (SA) is a highly controversial feature instituted in the name of national security by the DoD. Deliberate inaccuracies are introduced into the positioning information broadcast by the GPS satellites to non-military users who lack the capability to utilize the more accurate encrypted signals. It obviously cannot be a consistent inaccuracy or it would easily be calculated out by smart GPS user equipment. If you were to plot its variability on a graph, you might think of it as a "little old man" constantly wandering about the actual position at just a few miles per hour up to 50 or more meters in various directions. It also moves above and below the actual vertical position to the extent of about plus or minus 500 feet of elevation. GPS manufacturers have overcome SA's and other GPS inaccuracies within local areas, such as ports, airports, and survey sites, by developing a technique known as differential GPS. In simple terms, a GPS unit is surveyed in at a precisely known location. It's computer then determines the amount and direction of the error being introduced at any given moment. This corrective information is then transmitted to and utilized by other GPS units operating within the immediate area, which allows correction to be made by each unit. With differential GPS, accuracies to within less than a centimeter can be achieved with the use of highly sensitive equipment.

Then Secretary of Defense Aspin declared the system operational (January 1994). GPS is no longer experimental and can officially be used to guide navigation throughout the world. What does this all add up to for you, the land navigator? It provides you, absolutely free of charge, with a highly accurate three-dimensional position reporting capability with continuous, all-weather coverage available to an unlimited number of users anywhere on earth.

For a more in-depth technical exploration of GPS topics, we recommend contacting Navtech Seminars and GPS Supply in Arlington, Virginia at (703) 931-0500 or (800) 628-0885. The Navtech bookstore offers over 150 GPS related book and software titles. You may also access the Navtech Home Page at:

`http://nmaa.org/navtech.com/navtech.htm`

USGS: GEOGRAPHIC NAMES INFORMATION SYSTEM

APPENDIX C

In addition to providing large, medium and small scale topographic maps of the United States, as indicated in Chapter 3, the USGS National Mapping Program has developed an extensive data base of locations within the U.S. referenced by latitude/longitude grid pair coordinates. The data base which includes populated places, schools, harbors, reservoirs, tunnels, etc. is described by the USGS as follows:

"The Geographic Names Information System (GNIS), ... contains information about almost 2 million physical and cultural geographic features in the United States. The Federally recognized name of each feature described in the database is identified, and references are made to a features location by state, county, and geographic coordinates." See Figure C-1.

This source of information may be most useful for vehicle travel or when coordinates for the navigators destination are not readily obtainable from a regional topographic map.

The GNIS is accessible through Internet access at:

http://www-nmd.usgs.gov/www/gnis/gnis00.html

The entire database is also available on a compact disk entitled The Digital Gazetteer of the U.S. for the price of $57.00 by contacting any regional USGS Earth Science Information Center or phoning 1-800-USA-MAPS.

```
        Feature Name:Easton

              State:Pennsylvania

             County:Northampton

       Feature Type:populated place

          Elevation:300

 USGS 7.5'x7.5' Map:Easton

           Latitude:404118N

          Longitude:0751316W
```

FIGURE C-1:

RECORD FROM GNIS QUERY
FOR EASTON, PENNSYLVANIA

GLOSSARY
(TERMS & ACRONYMS)

Almanac	Data on the general location and health of all satellites in the GPS constellation. It can be collected from any available satellite.
Azimuth	A direction, usually measured clockwise in degrees, from a north reference line; sometimes referred to as a bearing. An azimuth can be referenced to true, grid, or magnetic north. Also, the reverse direction is referred to as a back azimuth.
Bearing	See azimuth.
CDI (Course Deviation Indicator)	The amount of deviation the GPS unit has accumulated (left or right) as its user negotiates the current leg of a route as measured in a perpendicular direction from the intended course line of the route.
Checkpoint	A natural or man-made feature that serves as a guide to movement and as the end of one route segment and the beginning of the next. GPS manufacturers are influenced by nautical navigation terms and often refer to checkpoints as waypoints.
COG (Course Over Ground)	The direction in which the GPS unit is moving with respect to the earth. COG can be reported in terms of either true or magnetic north values.

Contour Interval	The difference in elevation units between two adjacent index and supplementary contour lines. This information is generally reported in the margin of a topographic map.
Contour Line	A line on a map or chart connecting points of equal elevation. Most topographic maps today use four types of contour lines: index, intermediate, supplementary, and depression.
Default	The value or setting automatically chosen by the GPS unit.
ETA (Estimated Time of Arrival)	The time when the GPS unit should reach the destination of the current leg of the route based on the current speed of advance.
Functional Distance	A measure of distance that considers the time, effort, and level of difficulty involved in moving from one point on the ground to another.
Grid Reference Box	A box in the map's margin that contains instructions on how to determine grid coordinates and names the UTM MGRS grid zone designation(s) and the 100,000 Meter Square Identification(s) included on the map.
Grid Zone	One of 60 six-degree wide segments which circle the earth between the latitudes of 80° south and 84° north upon which the UTM grid is superimposed. The UTM grid is constructed parallel to the central meridian and the equator within each zone to form the coordinate base on large scale mapping projections. They are numbered from west to east beginning at the International Dateline (180° longitude).
Grid Zone Designation	The largest sub division of the six degree wide grid zones into six degree by eight degree segments used as the building blocks for the UTM MGRS grid coordinate format. These grid zone designations are numbered consecutively from west to east as grid zones and with alphabctical labels from south to north.

Latitude	The angular distance north or south of the equator measured by lines encircling the earth parallel to the equator in degrees from 0° to 90°.
Longitude	The angular distance east or west of the prime meridian (usually the Greenwich meridian) as measured by lines perpendicular to the parallels and converging at the poles from 0° to 180°.
Map	A graphic representation, usually on a plane surface and at an established scale, of natural and artificial features on the surface of a part or whole of the earth or other planetary body. Topographical maps show both the horizontal and vertical relationships of features portrayed, usually through use of contour lines.
Map Features	Four classifications of features are generally portrayed on maps: terrain, vegetation, hydrography (water), and culture (man made).
Map Datum	A method of assigning position coordinates to real world locations based upon an underlying ellipsoidal model of the earth used for drawing the map.
MGRS	The Military Grid Reference System alphanumeric format for reporting UTM grid coordinates. It is generally considered to be easier to use than the completely numeric UTM coordinate system.
One Hundred Thousand Meter (100,000) Square Identification	Each six degree by eight degree Grid Zone Designation is broken down into several dual letter identifiers, each 100 kilometers square, within the MGRS UTM coordinate format.
Position Fixing	Determining your position on a map by any means in terms of its coordinate system. If three dimensional, it will also include an elevation value above sea level.
Representative Fraction (RF)	The scale upon which a particular map is drawn (map distance as a numerator and ground distance as denominator).

SOA (Speed of Advance)	The component of the GPS unit's velocity which is in the direction of the actual destination.
SOG (Speed Over Ground)	The speed at which the GPS unit is moving with respect to the earth's surface.
SOSES	The five physical characteristics used for describing and identifying specific terrain features on the ground and on the map. They include: Shape, Orientation, Size, Elevation, Slope.
True Geographic Coordinates	Map coordinates reported in terms of latitude and longitude.
TTG (Time to Go)	The estimated amount of time needed for the GPS unit to reach the destination waypoint (checkpoint) of the current leg of a route based upon the current speed of advance (SOA).
Universal Coordinated Time (UT)	Universal Time, formerly referred to as Greenwich Mean Time (GMT).
UTM Grid Coordinates	The Universal Transverse Mercator Grid System used on most large and intermediate scale topographic and many other maps.
VMG (Velocity Made Good)	The component of the GPS unit's velocity which is parallel to the intended course line.
Waypoint	A position stored in the GPS unit's memory under a unique name (see Checkpoint).

ABOUT THE AUTHOR

NOEL J. HOTCHKISS

Noel J. Hotchkiss worked from 1985 to 1990 as a land navigation (LN) subject matter expert and instructional design consultant with the U. S. Army Research Institute (ARI) LN Training Team at the U.S. Army Infantry School, Fort Benning, GA. During the past three years, he has served as a consulting instructional program designer and contract LN instructor for Alexis International, Inc. on a project involving the military forces of the Kingdom of Saudi Arabia. In addition, he has written a two volume "how-to" LN manual based upon the Army's research findings. This book was published by Stocker & Yale, Inc., the manufacturer of military lensatic compasses.

While serving as a consultant to the U.S. Army ARI, Mr. Hotchkiss authored several LN research and training reports and programs as well as various articles appearing in professional military journals. In addition, he traveled to several service schools reviewing their

LN training in search of ways to improve the Army's overall performance in this crucial skill area. In testing some of the Army's new experimental LN training programs, he instructed soldiers enrolled in the 10th Mountain Division's "Light Fighter School" at Fort Drum, NY.

Mr. Hotchkiss earned an AB degree in social science, MA in social studies education, and MS in education administration at Syracuse University. He has also completed some graduate level study in the field of instructional design, development, and evaluation. His formal military education includes completion of the Armor Officer Basic Course (AOBC), Armor Officer Advance Course (AOAC-RC), several tactical intelligence officer courses, infantry and armor brigade staff officer refresher courses, a commanders' course, and Command and General Staff College Course (C&GSC-RC).

Commissioned through the R.O.T.C. program at Syracuse University, Lieutenant Colonel Hotchkiss' assignments included two years in the 2-63d Armor, 1st Infantry Division; eighteen years in the 27th Brigade, NY Army National Guard; and four years with the 1159th U.S. Army Reserve Forces School, 98th Division (Training). He has served as a tank platoon leader, battalion and brigade intelligence and operations officer, mechanized infantry battalion executive officer, and Area Commandant for officer and NCO educational system programs.

Since 1976, he has served as Principal at the Jordan-Elbridge Central High School near Syracuse, New York.

ACKNOWLEDGMENTS

First, I would like to thank the fine people at the Magellan Systems Corporation for their encouragement and assistance while writing this GPS Guide.

Richard L. Sill was particularly instrumental in encouraging Alexis USA, Inc. to undertake the development and publication of this book. In the area of Magellan's technical assistance, Stig Pedersen, Ed Lang, and Frank Houzvieka are to be especially commended for their patience in fielding those endless questions coming from me by letter and telephone as I completed the task.

I must also thank Wayne Gleason at Alexis USA, Inc. for asking me to write the book and having his fine training organization serve as its publisher. Also, Art Lipscomb again worked magic with his combined graphics and computer skills to accomplish the cover design, develop many of the illustrations, and layout the entire manuscript for publication. Alexis USA, Inc. specializes in the design, development, and delivery of training in several mediums, including interactive computer based training, and for many subjects, including land navigation.

Recognition must also be given to the fine research work done in the late 1980's by the Fort Benning Land Navigation Team of the U.S. Army Research Institute for the Behavioral and Social Sciences (ARI) for which I served as a consultant. I thank LTC Art Osborne (USA Ret.) at Litton Computer Services, Inc. and Dr. Seward Smith, Chief of the ARI Fort Benning Field Unit, for getting me involved in their effort to improve Army LN training.

I also wish to thank some of the outstanding people I have worked with over the years who taught me a great deal about training people in the subject of navigation. They include LTC Robert Berry, Commander of the 2-63rd Armor, 1st Infantry Division (Mech) (1964-65); SMAJ Mike Ratkoski and MSG Dick Thom, intelligence and training sections of the 27th Brigade, NY Army National Guard (1970-82); MSG Dave Mc Namara, 1159th U.S. Army Reserve Forces School (1982-86); LTC Dave Wooding, G3 of the 98th Division (Training), U.S. Army Reserve (1984-87); and Colonel Ron Jebavy (USA Ret.), Chief of the ARI LN Team.

Next, I recognize my debt to Louise (Weezie) Mullenix, Staff Editor at Litton Computer Services, Inc./ARI, who helped me often to sharpen my writing skills, and Mary Robb Mansfield, recently of the Jordan Elbridge High School English Department, who served to edit the manuscript of this publication. Nevertheless, all errors, whether in content, layout, or use of the English language, are to be attributed to me.

Finally, I wish to thank my full-time employer, the Jordan Elbridge Central School District, for their support and patience over the years as I flexibly planned and used much of my vacation time in the pursuit of my interests in training design, development, and instructional delivery; maps; and land navigation.

ALEXIS PUBLISHING

1037 Sterling Road • Suite 203 • Herndon, Virginia 22070

Please send ___ copies of *A Comprehensive Guide to Land Navigation with GPS* to:

Name _____

Street _____

City _____ State ___ Zip _____

Price: $29.95 X _____ copies $ _____

Sales Tax: Virginia residents add 4.5% $ _____

Shipping: $3.00/1st book, $1.00 ea. additional $ _____

Total Payment: $ _____

Form of Payment

☐ Check   ☐ VISA   ☐ Mastercard

Card # _____ Exp. Date: _____

Cardholder Name _____

Signature_____

Telephone Orders: (800) 200-8997
Fax Orders: (703) 318-7864

To order more copies of *A Comprehensive Guide to Land Navigation with GPS*, please remove (or photocopy) the above order form and return it, by fax or mail. Don't forget to ask about large order discounts.